DAMS AND RIVERS

A Primer on the Downstream Effects of Dams

By MICHAEL COLLIER, ROBERT H. WEBB, and JOHN C. SCHMIDT

U.S. GEOLOGICAL SURVEY
Circular 1126

Tucson, Arizona
June 1996

U.S. DEPARTMENT OF THE INTERIOR

BRUCE BABBITT, Secretary

U.S. GEOLOGICAL SURVEY

Gordon P. Eaton, Director

For additional information write to:

Chief, Branch of Regional Research, Western Region
U.S. Geological Survey
345 Middlefield Road, MS-472

Menlo Park, CA 94025

Copies of this report are available free from:

U.S. Geological Survey
Branch of Information Services
Box 25286
Denver, CO 80225

Foreword

The U.S. Geological Survey is charged with monitoring the water and mineral resources of the United States. Beginning in 1889, the Survey established a network of water gaging stations across most of the country's rivers; some also measured sediment content of the water. Consequently, we now have valuable long-term data with which to track water supply, sediment transport, and the occurrence of floods.

Many variables affect the flow of water from mountain brook to river delta. Some are short-term perturbations like summer thunderstorms. Others occur over a longer period of time, like the El Niños that might be separated by a decade or more. We think of these variables as natural occurrences, but humans have exerted some of the most important changes — water withdrawals for agriculture, inter-basin transfers, and especially the construction of an extensive system of dams.

Dams have altered the flow of many of the nation's rivers to meet societal needs. We expect floods to be contained. Irrigation is possible where deserts once existed. And water is released downstream not according to natural cycles but as dictated by a region's hour-by-hour needs for water or electricity. As a result, river channels below dams have changed dramatically. Depending on annual flow, flood peaks, and a river's sediment load, we might see changes ranging from sand building up in one channel, vegetation crowding into another, and extensive bank erosion in another.

This Circular explores the emerging scientific arena of change in rivers below dams. This science tries first to understand and then anticipate changes to river beds and banks, and to riparian habitats and animal communities. To some degree, these downstream changes can be influenced by specific strategies of dam management. Scientists and resource managers have a duty to assemble this information and present it without bias to the rest of society. Society can then more intelligently choose a balance between the benefits and adverse downstream effects of dams.

Robert M. Hirsch

DAMS AND RIVERS

A Primer on the Downstream Effects of Dams

U.S. GEOLOGICAL SURVEY

Circular 1126

A Primer on the Downstream Effects of Dams

By

Michael Collier, Robert H. Webb, and John C. Schmidt

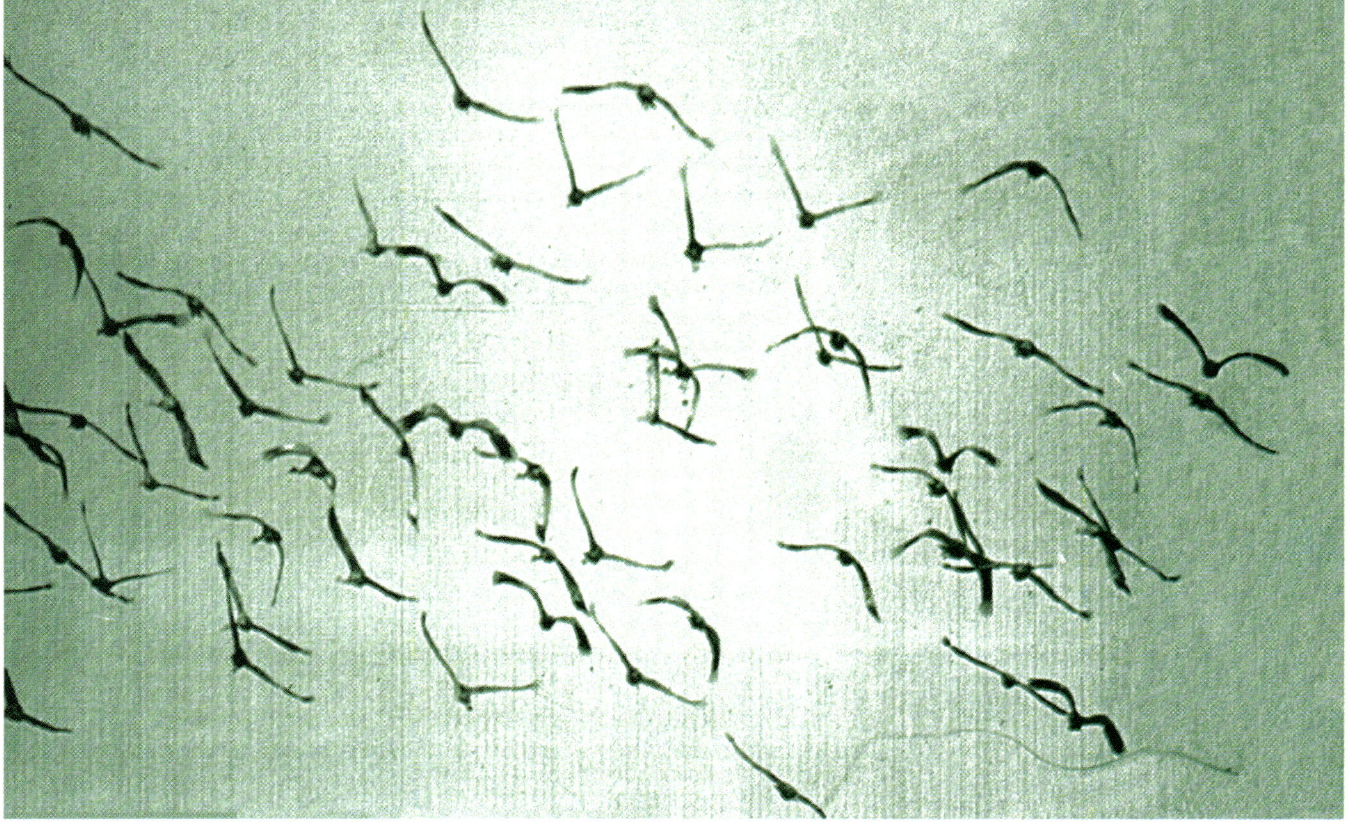

Table of Contents

Acknowledgements

We appreciate the input of Hector Morales, Lester Ray Talley, Frank Almendariz, John Taylor, and Doug Milam, who live and farm along the Rio Grande and Chattahoochee Rivers. The following individuals contributed to this project by discussing with the authors their involvement with, and observations about, dams along rivers: Curt Kennedy, Salt River Project, Phoenix; Woody Fine, Forest Service, Hells Canyon National Recreation Area; Dennis Womack, James Miller, Timothy Brewer, and Roger Gunn, Idaho Power Company, Boise; James Robinson and John Lee, International Boundary and Water Commission, El Paso and Presidio, Texas; Benjamin Everitt, Utah Department of Natural Resources, Salt Lake City; William Couch, Georgia Department of Game and Fish, Buford Fish Hatchery; Michael Thomas, Atlanta Regional Commission; Mork Winn, Georgia Department of Natural Resources, Atlanta; Michael Wilder and Gary Rakestraw, Georgia Power Company, Atlanta; Rich Valdez, Bio/West, Logan, Utah; Lloyd Greiner, Dave Sabo, Ron Moulton, and Suda Rajamani, Western Area Power Administration, Salt Lake City; Trevor Hughes, Utah State University, Logan; Gary Krapu, Henry Maddox, and Bob Williams, U.S. Fish and Wildlife Service; Gary Hendrix and Larry Stevens, National Park Service; Mike McAfee, Doug Young, Larry Crist, Bureau of Reclamation; and Edmund D. Andrews, Dallas Childers, Richard Marzolf, Jonathan Nelson, J. Dungan Smith, David Wangsness, and Garnett P. Williams, U.S. Geological Survey. Michael Collier took all of the photographs in this report. Moreover, Mike Yard and Dave Wegner, of the Bureau of Reclamation, and Brad Dimock, of the Grand Canyon River Guides, critically reviewed early drafts of the manuscript, whereas George Wise and George Dickel of the U.S. Geological Survey helped our thought processes both on the river and off. Designed and illustrated by Jeanne Sandra DiLeo. Our thanks to you all.

Conversion Factors

For readers who prefer to use metric units, conversion factors for the terms in this report are listed below

Multiply		by	To obtain
inch	(in.)	25.4	millimeter (mm)
foot	(ft)	0.3048	meter (m)
square foot	(ft^2)	0.09290	square meter (m^2)
cubic foot	(ft^3)	0.02832	cubic meter (m^3)
acre-feet	(ac-ft)	100233	cubic meter (m^3)
acres	(ac)	0.4047	hectare (ha)
mile	(mi)	1.609	kilometer (km)
square mile	(mi^2)	2.590	square kilometer (km^2)
cubic foot per second	(ft^3/s)	0.02832	cubic meter per second (m^3/s)
ounce avoirdupois	(oz avdp)	28.35	gram (g)
pound avoirdupois	(lb avdp)	0.4536	kilogram (kg)
tons, short	(2,000 lbs)	0.9072	megagram (Mg)
gallon	(gal)	3.785	liter (L)
degrees Fahrenheit	(°F)	$\frac{°F-32}{1.8}$	degrees Celsius (°C)

Sea level: In this report, "sea level" refers to the National Geodetic Vertical Datum of 1929 (NGVD of 1929)—a geodetic datum derived from a general adjustment of the first-order level nets of both the United States and Canada, formerly called "Sea Level Datum of 1929."

Maps

Figures

Introduction

Dams are an inextricable elemen of our society. We build dams fo multitude of reasons and at increasingly great cost. Scientist are beginning to examine some o the consequences of dam operation that were not fully appreciated in the past. Among the fore most of these are changes that can occur downstream. Some downstream effects may be mit gated by adjustments in dam

Glen Canyon Dam impounds the Colorado River near the border between Arizona and Utah

Dams and river regulation have become an integral part of our twentieth-century landscape and livelihood. Although untamed rivers are part of the cultural heritage of the United States, virtually every river in the lower 48 states is now regulated by dams, locks, or diversions. These regulated rivers have afforded society many benefits — cheap electricity, navigable streams, absence of devastating floods, decreased threat of drought. But regulated rivers are fundamentally different ecological and physical entities from untamed rivers. Natural cycles of flooding and sediment transport have been eliminated from many of these rivers. Channel shape, riverine vegetation, and in-stream aquatic communities have, in many cases, changed as a result of flow regulation.

Because of the scale of dam construction that has taken place in the United States, society now has before it a set of choices regarding the kind of river characteristics we desire. Like it or not, we control the destiny of these streams. Traditionally, river managers have focused on issues of engineering efficiency, sometimes to the neglect of in-stream environmental values. The engineering matters remain a focus of management, but our society must also choose whether or not to manage rivers for their intrinsic environmental values. We can consciously choose to manage our rivers for certain anticipated environmental consequences, or we can intentionally choose to accept the environmental responses as they haphazardly occur. The purpose of this Circular is to illustrate the downstream environmental consequences of dams, and to explain the basis on which rivers can be scientifically managed.

Egyptians were building dams upstream from Cairo 5,000 years ago. Western Europeans constructed dams to power water wheels during the late Middle Ages (Smith, 1971). Eight hundred years ago, the Anasazi built small check dams on Mesa Verde in Colorado to hold storm runoff for later use on their crops (Ortiz, 1979). As early as AD 833, the Chinese used human and animal power to build a 90-foot-high dam on the Abang Xi River (Petts, 1984). Twelve centuries later, this dam is still used for irrigation diversion although the reservoir has filled with sediment.

But the age of widespread, large-structure dam-building awaited the arrival of heavy machinery and the high ambitions of industrialized societies. In the United States, the pace of large dam construction hit its quickest stride between 1935 and 1965 (Thomas, 1976). In the West, Hoover and Grand Coulee were completed before World War II; Glen Canyon Dam was finished in 1963. In the East, creation of the 26-dam Tennessee Valley Authority system ushered in an era of building dams and managing reservoirs integrated over an entire basin (Cullen, 1962). Currently, there are more than 75,000 dams higher than six feet in the United States; the reservoirs behind these dams cover about 3 percent of the nation's land surface (R.F. Stallard, oral commun., 1994). Worldwide, 193,500 square miles (mi^2) of land is inundated by reservoirs. Now in a given year, 60 percent of the United States' entire river flow can be stored behind dams (Hirsch and others, 1990). In the dry American Southwest, dams on the Colorado River can store four years of typical flow (Andrews, 1991). More dams are being built in developing countries, but in the United States, Canada, and Western Europe only a few potential sites for large dams remain under realistic consideration.

Imperial Dam on the Colorado River upstream from Yuma, Arizona

Have we benefitted by building these dams? Viewed in one dimension, the answer is a resounding "Yes." Hydroelectric powerplants harnessing the Columbia River and its tributaries produce 75 percent of the American Northwest's electricity (Palmer, 1991). Each year, 8.2 million acre-feet of water are diverted from the lower Colorado River to homes and farms in California, Arizona and Mexico through aqueducts that cross hundreds of miles of intervening desert. None reaches the gulf of California. Since dams were built across rivers in the Connecticut River Valley, no floods have occurred like the ones that killed 108 people and crippled the towns of Bolton and Hartford in 1927 and 1936 (Leuchtenburg, 1953).

In the simplest sense, we build dams for the same reason we wear coats in the winter: to exert control over an aspect of an environment that would otherwise make living difficult or even impossible. If a valley is subject to destructive flooding, we dam its river. If the desert is dry, we build a lake. The list of available reasons for building a dam is long and complex – for water storage to quench municipal, agricultural, and industrial thirsts; for flood control and improved navigation; for sediment trapping; for improvement of water quality; for electrical power generation; for recreation, aesthetic, and wildlife considerations.

As dams became bigger and more expensive, a wider array of benefits was needed to justify the cost of dam construction. Most dams built after 1950 had many purposes, and sometimes these purposes were in competition with each other. Glen Canyon Dam in Arizona was initially conceived as a tool to balance the water allocations between the upper and lower basin states of the Colorado River (Ingram and others, 1991). But with an initial price tag of $325 million, construction of the dam needed additional justifications. Water conservation, downstream distribution, and hydroelectric power were written into the dam's operation considerations. Recreation and flood control subsequently were added to the dam's operating criteria.

Environmental Values Downstream from Dams

Downstream effects of dams were of little concern during the design and construction of most dams in the United States. Engineers knew that clear water releases would erode the channel immediately downstream from spillways and power plants; they attempted to calculate the amount of scour to protect the integrity of the dam and its structures. Changes in fish populations or riparian vegetation were often unanticipated or were not taken seriously: in fact, to build a game-fishing industry, some channels downstream from dams and upstream from reservoirs were poisoned to remove native fish. It took a little known, and endangered, native fish downstream from the proposed Tellico Dam in Tennessee to focus attention in the 1970s on environmental changes associated with dam operations.

Society values its rivers in a more complex manner in the 1990s than it did when most dams were originally constructed. More than ever; we count on the traditional products of dam operation – water, power, flood control. But we are increasingly aware of the environmental values of our rivers. We enjoy whitewater recreation in an increasing fashion; many classic whitewater boating trips are downstream from dams. We flock to the blue-ribbon trout fisheries that thrive in the tailwaters of dams. We treasure wilderness, and many regulated rivers flow through wild and inspirational landscapes. As a society, we have insisted on some protection of our rivers, and the Wild and Scenic River designation, enacted by Congress, has been one result.

Several issues take the forefront in consideration of adverse effects of dam operations. Native fish, protected under the Endangered Species Act are thought to be threatened by the clear, and usually cold, releases from dams. Riparian vegetation can either be enhanced or degraded by dam operations. Streamside and channel sedimentary deposits are critical: too much sediment can aggrade channels and cause flooding problems, whereas erosion of sediment can degrade habitat and decrease the potential for recreational use. These issues are among those that may drive change in the way our dams are operated.

Kayakers on the Chattahoochee River at Duluth, Georgia

To the uninitiated, a dam might seem little more sophisticated than a plug stuffed into a pipe, a few shovelfuls of dirt thrown across a ditch. But a deeper look brings an entire world of technological expertise into view. Engineers must identify appropriate sites for a dam, locate materials for its construction, conceive the basic shape of the dam, decide whether or not to build an associated hydroelectric power plant, and calculate the necessary size of emergency spillways. The design of a dam is directly tied to its fundamental purpose. Run-of-the-river designs are usually low in elevation, have small upstream reservoirs, and modify the natural flood and sediment-transport cycle only slightly. Alternatively, high dams with large upstream reservoirs can store many months, if not years, of natural streamflow and can generate prodigious amounts of hydroelectric energy due to the fall of the river at the dam.

Dams and reservoirs differ not only in their sizes, but also in operational strategies. Dams of the same size may hold varying amounts of water depending on their ultimate purpose. A flood control dam keeps its reservoir low at the onset of each flood season, while a water-supply reservoir tries to remain full as long as possible. Because the actual inflow into a reservoir can never be precisely anticipated, each of these operating strategies carries inherent risks that the reservoir will overflow or that the reservoir will go dry. The task of the water-resource engineer is to develop techniques that will increase the chances of achieving the desired objectives of the dam, while accepting some risk that other situations, less relevant to the dam's ultimate purpose, will occur. In other words, a flood control dam has some chance of going dry and a water-supply dam has some chance of spilling.

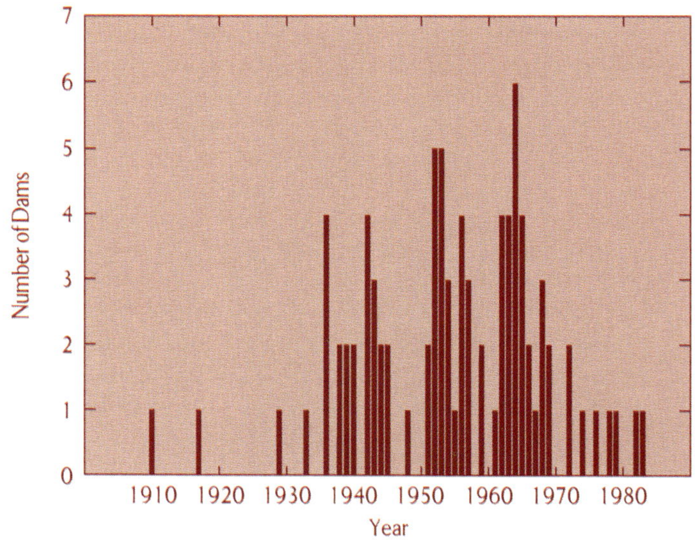

Dates of closure for dams in the continental United States with reservoir capacity of a million acre-feet or more (compiled by Graf, 1993).

But we can't have it all, all of the time. The challenge for today's manager is that yet other factors now exist in the dam-management equation – those centered on management of the downstream river corridor and its ecology. Scientifically-based management of regulated rivers adds a new layer of objectives to the already complex task of the water-resource engineer who designs the strategies of multi-purpose reservoir management.

The halcyon days of building large dams in the United States have passed. Most of the "good sites," based purely on the engineering perspective of high canyon walls, solid foundation, and a large basin upstream, have already been used. The price tag on a dam is now orders of magnitude greater than for equivalent structures built during the 1930s. But more importantly, the American public has grown to expect a full cost/benefit accounting of a large project, not just in terms of construction cost versus immediate benefits, but also

The ratio of reservoir storage to annual water supply in parts of North America (modified from Hirsch and others, 1990). The western United States and southwestern Canada have the most extensive reservoir development relative to available water supply. More dams have been constructed in the Columbia River basin and in the Tennessee Valley (neither are shown), but these basins have higher water yields. Many of the largest problems with downstream effects of reservoirs are in the basins with the highest ratio of reservoir storage to annual water supply because dam operations are constrained more in those basins.

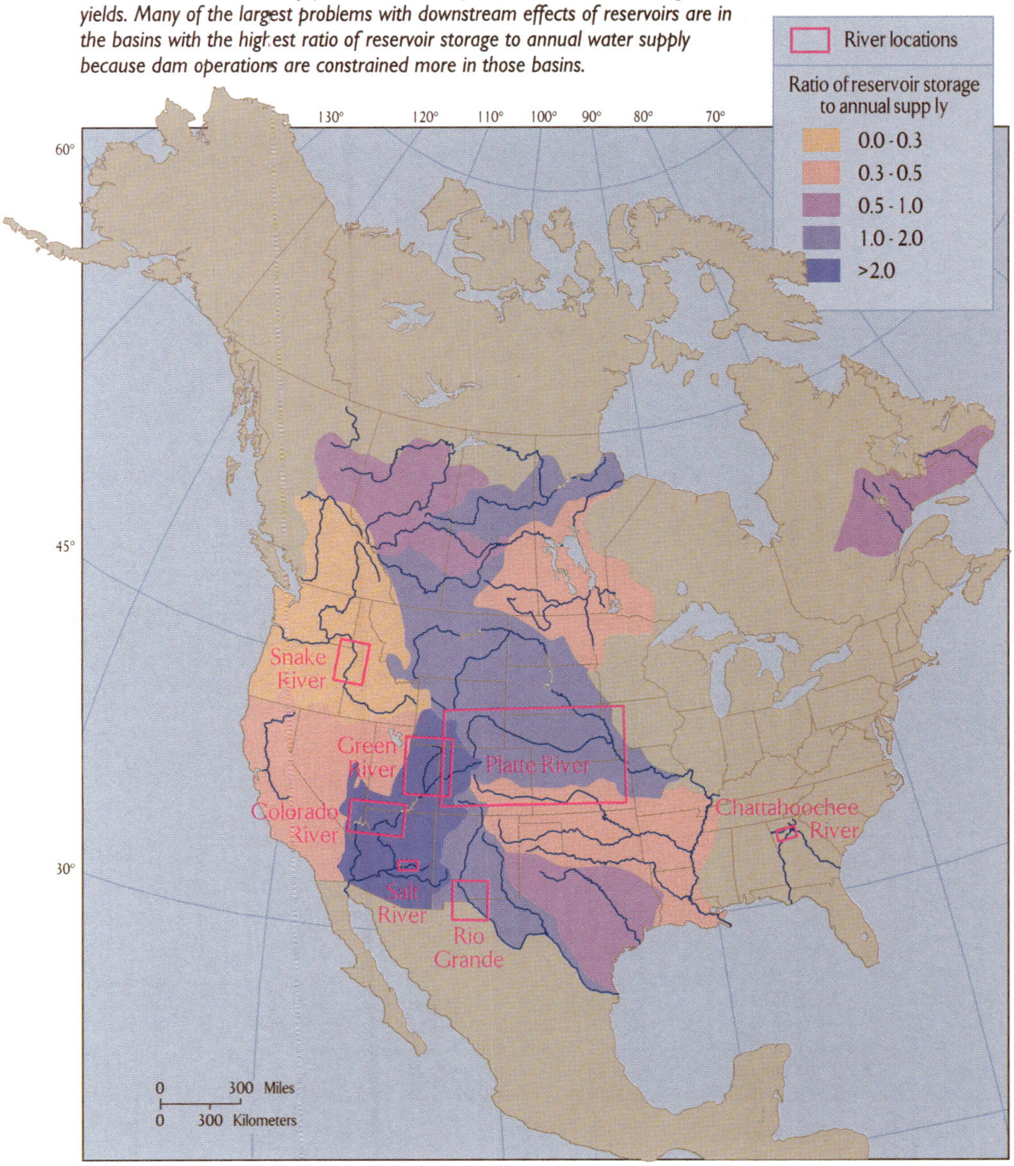

Diversion dam on the Colorado River near Blythe, California

The Rio Grande above Mesilla Dam near Las Cruces, New Mexico

in terms of long-range environmental and social costs. Instead of building new dams, we are now spending a lot of time, energy, and money examining the effects of existing ones. The question now relevant to dams in the United States is not "Should we build another one?" but "How can we best use the ones we have?"

Historically, our society has always found it easier to build than destroy a dam. Driven by the memory of a simpler time and an unfettered river, some people are fond of contemplating the elimination of certain dams. In a few instances, the environmental costs of an old dam are deemed so great that the dam's removal is conceivably warranted. Examples include the Edwards Dam on the Kennebec River in Maine, the Elwha and Glines Canyon Dam in Olympic National Park, and Hetch Hetchy Dam within Yosemite National Park in California. But by and large, this is rarely a realistic option.

Once a dam has been built, we reap its benefits and learn to live with the environmental effects. The real question then becomes: can a dam be operated so as to maximize its benefits and minimize its costs? The exciting answer is "maybe." The ground rules for answering this question involve taking some long hard looks at costs and benefits. What benefits do we value, what costs are we willing to bear? What new values — like downstream ecological impact — have been brought to bear in this accounting since the dam was designed and built? Arriving at a meaningful answer requires setting aside political and personal biases long enough to honestly say what we want from a dam and its river, and to accurately evaluate all the dimensions of impact that a dam can have. Society must decide what it wants; scientists can help to show what we are likely to get under different management strategies.

Viewed in one carefully chosen dimension, many dams have been worthwhile — this dam prevented flooding, that dam generated a lot of electricity. But with time, we have also come to realize that the adverse environmental effects of a dam may extend in circles far wider than had been appreciated in the past. For decades, people have known and argued about the more obvious effects of dams: flooded valleys and displaced farmers; fish migrations blocked or disrupted; one state taking water needed by another state downstream; water quality improved or impaired. We did not spend a lot of time thinking about the issue of downstream effects when conceiving dams during the first half of the twentieth century. But in the past twenty years, scientists and the public have begun to appreciate an additional effect of dams: changes to the downstream river environment.

The river emerging from a dam is not the same river that entered its reservoir. That new river may be hotter or colder. Its daily discharge may vary wildly, while its seasonal pattern of high spring floods and low winter flow may be inhibited beyond recognition. Suddenly starved of its sediment load, the clear waters of a river below a dam may scour its bed and banks. An entirely new succession of riparian plants and animals may move into the river and valley below a dam. Native fishes may die or be severely stressed.

Oxbow Dam, part of the Hells Canyon Complex, on the Snake River, Idaho

Removing Dams

The Elwha River, which drains part of Olympic National Park in northwest Washington, had a significant salmon spawning run before it was impounded in the early 1900s. Elwha Dam, 110 feet high, was built in 1913 and forms Aldwell Lake; Glines Canyon Dam, 185 feet high, was built in 1926 and forms Lake Mills a short distance upstream from Elwha Dam. The dams were built by a logging company to supply electrical power to a wood products plant at Port Angeles; the city currently draws its municipal water supply from Lake Mills. Many years after their construction, Olympic National Park was established on land that included the dams.

The National Park Service and other Federal and State agencies are concerned that the dams block spawning runs of endangered salmon up the Elwha River. The pre-dam salmon runs on the Elwha were large and famous. Because the livelihood of these fish and their cultural importance to local tribes are considered more important than the economic benefits of the dams and reservoirs, the National Park Service is planning to dismantle Elwha and Glines Canyon Dams and return the Elwha River to its original condition. One significant problem with restoration is management of the sediment in the reservoirs; extensive deltaic deposits in the upper ends of both lakes must be redistributed in a fashion to minimize downstream effects once the dams are completely dismantled. Estimated costs for dismantling the dams are between $60 and $200 million; costs would be lower if the river is allowed to redistribute the sediment instead of solutions involving physical removal of the sediment. The cost of dismantling these dams will be borne by U.S. taxpayers.

Glines Canyon Dam on the Elwha River in Olympic National Park, Washington

During the past two decades, earth scientists have become increasingly involved in the study of downstream effects of dams. The U.S. Geological Survey has maintained a nationwide network of stream gaging stations since the late 1800s. Some stations have records of the amount of sediment transported downstream. These records provide invaluable data concerning the behavior of a river before and after being dammed.

In this Circular, we explore the downstream effects of dams. First, we look at a free-flowing river – the upper Salt River of Arizona – and its natural cycles of flow and sedimentation. Then we examine six regulated rivers; the Snake, Rio Grande, Chattahoochee, Platte, Green and Colorado Rivers. Each of these rivers highlights a particular use of a dam or a particular downstream effect. Finally we discuss the role of science in managing dams.

Upper Salt River

A Natural Stream

The upper Salt River in central Arizona is unregulated. Flow and sediment transport vary widely from one season to another, and from year to year. Occasional high-magnitude floods move cobbles and boulders into bars, which form rapids that attract recreationalists. Native riparian vegetation forms a dense band above flood level, and nonnative vegetation is relatively sparse.

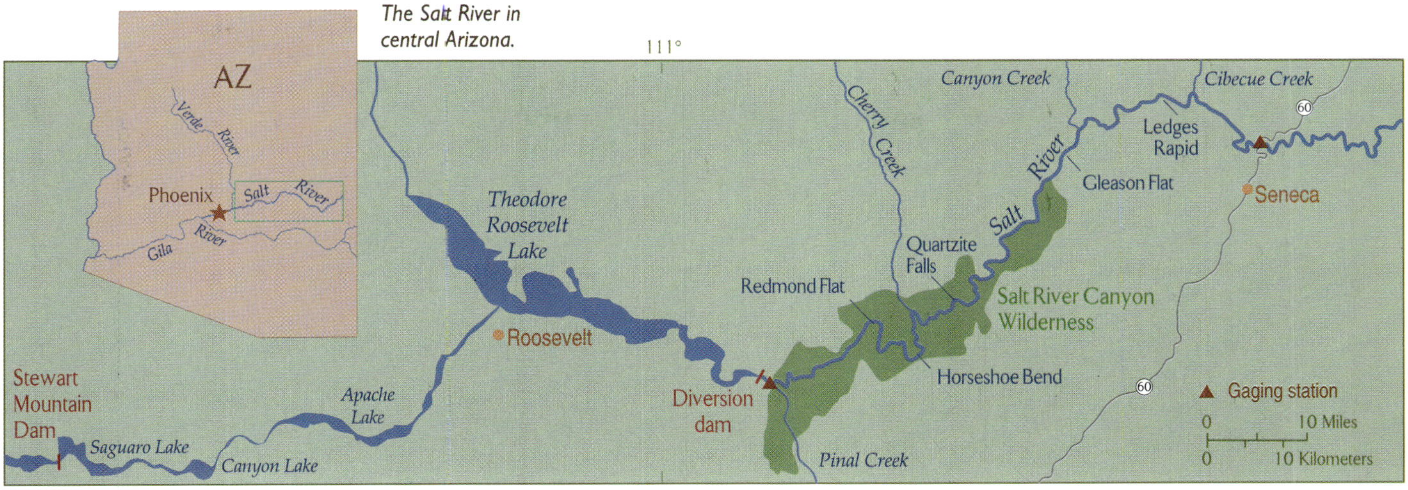

The Salt River in central Arizona.

The upper Salt River is a boisterous stream tumbling out of the White Mountains of eastern Arizona, west towards its confluence with the Gila River near Phoenix. The first 125 miles of the Salt, unimpeded by dams, are good examples of an unregulated river. The Salt River Canyon, is a major east-west trending gash in central Arizona. Downstream the river's character changes dramatically. Beginning at Roosevelt Lake reservoir, the Salt becomes a "working" river. A series of four reservoirs that supply water for irrigation, industrial, and municipal use in Phoenix marks the end of the free-flowing river.

A natural river has an annual cycle of floods and low-flow periods, depending on climate and season. The upper Salt River basin receives moisture from several types of storms. The greatest amounts of moisture are delivered to the basin by fall and winter storms that roll in from the Pacific Ocean, this moisture is stored as snow pack in the high country. If warm rain falls on previously fallen snow, the Salt River can suddenly snap from trickle to torrent. On December 17, 1978 a severe winter storm dumped 10 in. of rain on parts of the Salt River drainage (Aldridge and Hales, 1984). Before the storm, the Salt River was flowing at 3,460 cubic feet per second (ft^3/s), twenty-two hours later the river surged to 95,000 ft^3/s. Prehistoric floods have reached 160,000 ft^3/s (Partridge and Baker, 1987), yet flow of the Salt is sometimes less than 100 ft^3/s in early summer. This river may be an extreme example of variability in discharge, but many western rivers probably rose and fell just as quickly before they were dammed.

Big floods, little floods. Wet years, dry years. So what? Geoscientists must put all these numbers into perspective when they try to understand how a river works. They study basin and river geometry in order to understand how floods sculpt the river's natural channel. Hydrologists try to analyze historical patterns of flow in order to anticipate floods that can be expected in the future. Biologists study aquatic organisms and riparian plants in relation to these patterns of flow.

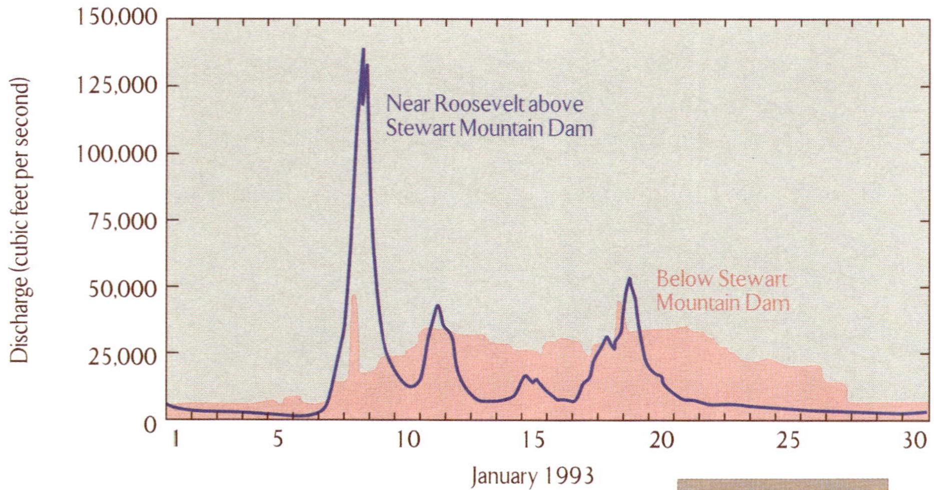

Hydrograph of the Salt River for January 1993. Peak discharge reached 143,000 ft^3/s on January 9 above Roosevelt Dam. Releases from Stewart Mountain Dam, the farthest downstream of four dams on the Salt, are dampened and protracted compared to the inflow to Roosevelt Lake.

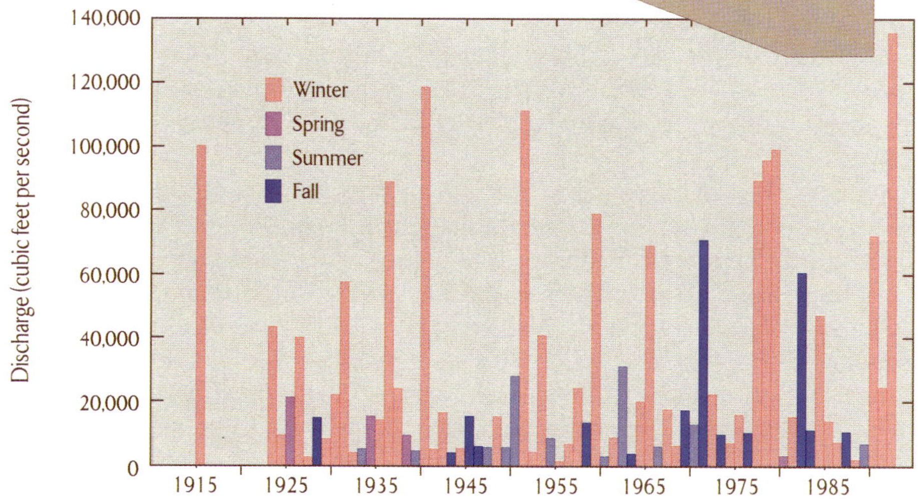

Annual flood series for the Salt River. The large variability in annual floods is an environmental stress to which native riparian vegetation and fishes have adapted.

Unregulated rivers in the western United States carry prodigious amounts of sediment during floods. The Salt River is no exception. The sediment load carried by the Salt can vary by orders of magnitude. Much of the year, the river is relatively clear. But in a large flood, the Salt can carry extremely large quantities of sediment, scouring material from the bottom and sides of its channel that normally is not disturbed by the flow of the river.

The amount of sediment transported by a river, termed its capacity, increases dramatically with discharge, typically in an exponential fashion. That sediment may be suspended in the water, or bounced along the bed. The river's competence to pluck larger and larger particles from its bed is a direct function of water velocity, as well as the shape of the channel where a particular sand grain, cobble, or boulder happens to lie.

We think of rivers as the cutting edge of erosion. When viewed through the long lens of a geologic perspective, it is true: rivers do carve their valleys. But when viewed through the shorter focal length of an historical perspective, a river typically is in dynamic equilibrium with its valley. Sometimes the river scours its bed during floods or high flows of long

The Salt River at Canyon Creek

duration; sometimes it builds up (aggrades) its bed during lower-flow periods as sidestreams continue to deliver large quantities of sediment. For a particular segment of river channel, the instantaneous elements of this equilibrium include discharge, sediment load, and channel slope, shape, and roughness (Pickup, 1976). At any instant, a river will adjust its channel to these elements. Each river system displays a unique response to the inputs of these elements (Baker, 1977). Confronted by floods of equal magnitude, a segment of river confined to a bedrock channel will behave quite differently than another segment flowing across an alluvial plain.

Riparian vegetation plays an essential role in the evolution of a river corridor. Streamside trees and brush take advantage of perennial surface water and fine-grained substrate for growth. The plants may be native – willows, cottonwoods, and mesquite – or exotics.

Tamarisk, a salt-tolerant brushy tree introduced to the West sometime in the late 1800s (Robinson, 1965), is the most common non-native species. Once established, vegetation can directly influence the impact of high flows by increasing channel roughness and decreasing flow velocities (Graf, 1978).

The Salt is a tributary of the Gila River, Arizona's largest water course. Much of the Gila's course is lined with tamarisk. Burkham (1972a) did a detailed study of the Gila in the 1960s and 1970s. Moderate-sized floods were observed on three occasions, before and after dense thickets of tamarisk were intentionally removed from the banks of the river. Not surprisingly, when the channel was clogged with tamarisk, it could not efficiently handle even moderate flood. As the water sieved through the brush, the river slowed and dropped its load of sediment. The floodwaters rose higher above the river banks and spread farther out from the banks where the channel was obstructed by tamarisk. Since Burkham's work, large floods on the Gila in 1972, 1978 to 1979, 1983, and 1993 have ripped most - but not all - of the tamarisk from the floodplains.

Mesquite thickets, called bosques, are common along the Salt and Gila Rivers (Minckley and Brown, 1982). Mesquite uses less water than other species, such as tamarisk, and can grow at higher elevations above the level of the river. Mesquite also can grow

Kayaker in Ledges Rapid on the Salt River

on rocky substrate in addition to sandy floodplains. Periods of small floods, such as the middle of the 20th century (Webb and Betancourt, 1992), allow bosques to encroach upon the floodplain. As with tamarisk thickets, floods wreak havoc on mesquite bosques. In one channel bend between the bedrock walls of the canyon, floods on the Gila between 1972 and 1979 halved the size of one bosque (Minckley and Clark, 1984). Most of the destruction was caused by lateral erosion of the channel banks, as is the case with tamarisk, but mesquites can better withstand flooding because of their toehold in the rocky slopes.

The Value of Streamflow Gaging

Hydrologists have used streamflow gaging stations for more than a century to assess the water resources in the United States. The first gaging station was established on the Rio Grande River near Embudo, New Mexico, in January 1889. Currently, the U.S. Geological Survey operates about 7000 gaging stations in the 50 states, Puerto Rico, and islands in the Pacific Ocean. Although the early gaging stations were read daily by operators who lived nearby, modern gaging stations can transmit data instantaneously to satellites and ground stations as much as a thousand miles away. Gaging stations provide the necessary data to plan water development and use, to manage water resources to meet specific legal requirements, to evaluate the effect of climate and land management on water supply, and to quantify the relation between flow and environmental quality of a river.

The amount and variability of streamflow is the single most important feature to monitor in a regulated river. These data are needed to quantitatively evaluate the impact of flow regulation on a river. Most regulated rivers in the United States were gaged before dams were built; comparison of flow before and after regulation provides insight into the reasons for changes brought about by dam operations. Even for regulated rivers where streamflow was not recorded before regulation, establishment of new gaging stations is essential to monitor the ongoing changes brought about by dam releases.

Native fish evolved under the extreme variability of the unregulated river. The Salt and Gila Rivers are home to 20 species of native fish (Minckley and Brown, 1982), most of which are endangered as a result of water development and introduction of game species elsewhere in the lower Colorado River drainage. The temperature of the Salt can vary from near freezing in winter to 77 degrees Fahrenheit in summer; any fish within the river must be prepared to deal with these extremes. The fish also must contend with environmental factors such as extreme changes in flow and large variations in sediment concentration.

Three views of Hess Creek at the Salt River

The largest flood ever recorded (143,000 ft^3/s) rolled down the Salt River into Roosevelt Lake on January 8, 1993. Tamarisks and willows were stripped from reaches where they had gradually encroached during years of lower flow, mesquite bosques were damaged but not destroyed. The native fish probably were

little affected. Certain cobble bars were scoured away. Some banks were severely eroded. But this river, like other natural rivers, gives back almost as much as it takes away. Where the river is confined by bedrock within the Salt River Canyon, tremendous quantities of new sand were deposited within eddies and along the channel margins. Clean new beaches remained perched 40 feet above the river after the flood receded. Cobble bars and boulder fans within the channel were reworked and in many places, built up. Roosevelt Lake acts as the upper Salt's temporary base level, and farmers along the lake's margin were pleased to find new topsoil in their fields after the floods receded. In the days prior to construction of Roosevelt Dam, such a flood would have inundated a broad swath of the desert where Phoenix now exists; terraces of sediment would have been deposited throughout that floodplain.

The upper Salt River has all the characteristics of a healthy unregulated river. Typical of rivers in the Southwest, it displays a wide range of flow and sediment transport – it is capable of quickly changing from minimal flow to awesome flood. This river is always in the process of adjusting its channel to the equilibrium that exists between erosion and deposition. The Salt offers a standard against which to compare regulated rivers.

Snake River

Hydroelectric Power, Channel Degradation

Hydroelectric power is one of the most important benefits of river regulation. Some large dams generate base-load power, but dams with limited reservoir storage typically are used to provide power at times of peak demand. The Hells Canyon Complex of dams on the Snake River illustrate a reservoir system where hydroelectric power is the overriding priority. The generation of hydroelectric power has severely changed the normal dynamics of the river's flows. Dams on the Snake block historic salmon spawning runs, and frequent high releases have caused depletion of sand downstream from the dams.

The Snake River is the most extensively dammed river in the West. Twenty-five dams lie between its headwaters in Yellowstone National Park and its confluence with the Columbia River, 1,000 miles downstream. The Snake is one of the major tributaries of the Columbia River, which drains 259,000 mi^2 of Canada and the Pacific Northwest. Starting with the completion of Grand Coulee Dam, the Columbia has effectively been converted from 1,210 miles of free-flowing river (Bartlett, 1984) to a series of dams and reservoirs, each backing into the next, capable of generating far more electricity than the Northwest was initially able to use. The aluminum industry,

Brownlee Dam

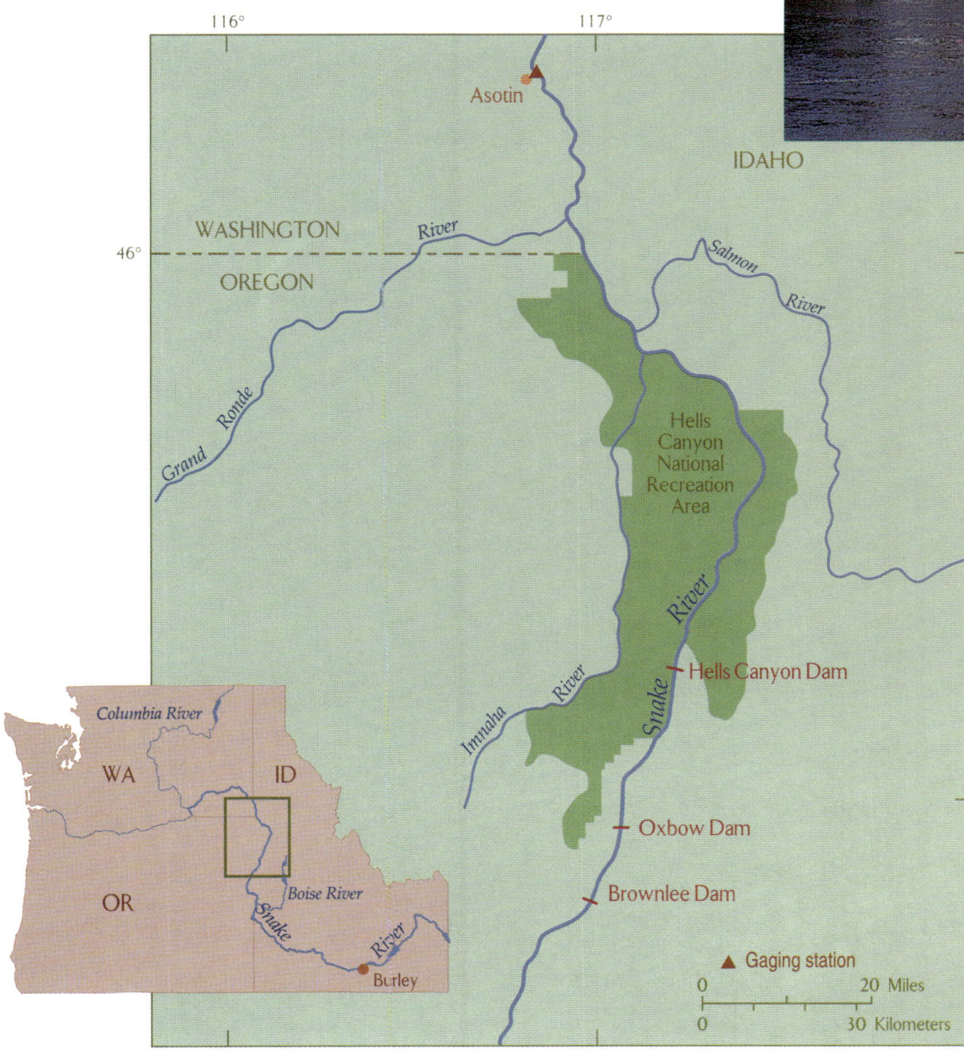

The Snake River in Oregon, Idaho, and Washington

with its insatiable appetite for electrical power, was invited to move into the Northwest in the 1940s and 1950s after the dams were built. In this case, demand for power followed supply.

The Snake River is many things to many people. Idaho farmers like to think of the Snake as a "working river." Halfway between the Snake River's headwaters and the Columbia, diversions at Milner Dam near Burley, Idaho deplete all but a trickle (200 ft^3/s) of the river's flow. This water irrigates more than three million acres of farm land, an area roughly

The Snake River in Hells Canyon

the size of Connecticut (Palmer, 1991). That's a lot of potatoes. Tim Palmer recalls his first impression when he visited Shoshone Falls, now usually dry, an impression that stands in sharp contrast to the farmers who depend on Snake River water for irrigation: "I wondered where the water had gone, and stood puzzled, feeling that nature had been warped in a sinister way, as if I had seen a three-legged deer or a toothless squirrel."

After passing west through Idaho, the Snake swings north to outline the Idaho/Oregon border. Below Milner Dam, the Snake is recharged by the Thousands Springs (whose source is in part the return flow from all those potato fields) and then by the Boise, Owyhee, and Payette Rivers. The Snake once again is full-blooded as it rolls into Hells Canyon with a yearly discharge of 16 million acre-feet. The river drops into canyons that make farming and even ranching progressively more difficult. Hells Canyon is arguably the deepest canyon in the United States. Peaks loom 7,900 feet above its waters.

Deep canyons and a big river were the siren's song that few dam builders could resist.

In 1906, the Idaho-Oregon Light and Power Company tried to take advantage of the topography at Oxbow, by building a dam and then drilling a 1,000-foot tunnel to shortcut a two-mile loop of river (Carrey and others, 1979). Despite a significant investment, the company managed to generate only 600 kilowatts (kw) at this facility; with debts mounting, the company filed for bankruptcy. Later during the 1940s, the U.S. Army Corps of

Brownlee Dam on the Snake River

Engineers and the Bureau of Reclamation were maneuvering for the rights to build a 590-foot dam in Hells Canyon. But the Idaho Power Company, reorganized from the ashes of the Idaho-Oregon Light and Power Company, took advantage of options that it had inherited at the Oxbow, and received a license to dam the Snake River upstream at Brownlee. Short-circuited by Idaho Power Company, the two Federal agencies would have to be content with their eight dams further downstream on the Snake and the Columbia. In all, Idaho Power built three dams, called the Hells Canyon Complex, within a 35-mile stretch of the Snake. Brownlee was completed in 1958, a new Oxbow Dam in 1961, and Hells Canyon Dam in 1967.

When the gates were first closed, combined storage of the Hells Canyon Complex was one million acre-feet of water, with 90 percent held in Brownlee Reservoir. All together, that represents only seven percent of the river's average annual flow as measured at Hells Canyon. Contrast that capacity with storage behind Glen Canyon Dam on the Colorado River, where the single reservoir holds 2.3 years of that river's flow. Because the Hells Canyon Complex doesn't have much storage capacity, the dams have little value for flood control, and managers are able to maximize the potential for the generation of electricity.

Idaho Power Company is tied into a grid that provides electricity throughout the West. The demand for electricity within this grid is not uniform. Daily peak demands occur during summer afternoons or early winter mornings. The need for power drops off markedly on weekends. Weather conditions (and consequently the power needed for

Obstruction of Fish Runs

Salmon once crowded the waters of the Snake River but are increasingly rare these days. Summer and fall chinook and sockeye salmon would migrate up from the Pacific Ocean through the Columbia and Snake Rivers to spawn on the tributary of their origin. Five to fourteen percent of adult salmon are killed at each of the eight dams through which they pass on their way up the Columbia (Eley and Watkins, 1991). When spawning is successful, the young fish have even lower rates of success in migrating downstream through the reservoirs. The chinook are now listed as threatened species, and the sockeye salmon is considered endangered (Stuebner, 1993). Idaho Power Company built fish ladders and other bypass systems into each of its dams of the Hells Canyon Complex, but all were unsuccessful. Now no salmon migrate above Hells Canyon Dam.

A few remaining salmon struggle each year to reproduce below Hells Canyon Dam. Operations of the Hells Canyon dams (not to mention all the other dams on the Columbia River system) have catastrophically affected salmon habitat. Will the fish hatcheries that the power company has funded be able to sustain even a small percentage of this fish population? At this point, it is hard to be optimistic for the fish unless the dams are removed or more effective measures are taken to move fish around the dams.

More than most, the Hells Canyon Complex illustrates the collision of hydroelectric-power generation and environmental values — and how complicated (and potentially expensive) solutions may be.

Hells Canyon Dam on the Snake River

heating and cooling) can vary sharply from one part of the grid to another. Instantaneous changes can occur if a key transmission line suddenly drops out of service. Power companies must "wheel" electricity from one region to another, and the entire grid must be able to instantly respond to sudden fluctuations in demand.

Throughout the West, a large proportion of electricity is generated by coal-fired and nuclear plants that efficiently supply constant maximum levels of power. These thermal plants become very inefficient, however, when they are run at less than maximum capacity. Once shut down, these plants take hours to come up to full steam. Electrical utilities are better off buying additional electrical power from another utility at a premium price to cover brief peak demands, rather than covering peak demands by investing in additional coal-fired plants or natural gas turbines that will be used for only a few hours a day.

Hydroelectric power, on the other hand, can be brought on line in a matter of minutes. Turbine efficiency remains high throughout a wide range of dam releases. Consequently, hydroelectric power has long been viewed as an ideal asset with which to respond to perturbations of demand within a power grid. This ability to instantly generate more power is valuable, and "peak power" is sold for considerably more than power generated during "off-peak" or base-load periods. Idaho Power Company operates coal-fired generating stations that supply base-load energy, but the company obtains all of its peak power from dams of the Hells Canyon Complex. The company tries to hold the water of the Snake River behind the dams when electrical demand is low, and releases water when demand (and the price per kilowatt - hour) is high.

The Hells Canyon Complex has the capacity to generate 1,400 megawatts (mw) when releasing 30,000 ft^3/s from all three dams. Larger discharges must flow through spillways, bypassing both the dams' generators and the company's revenues. More than half of the total generation capacity comes from Brownlee Dam. When possible, the company follows the fluctuations of power demand with its Brownlee units. Water released from Brownlee quickly passes through the small Oxbow reservoir into Hells Canyon Reservoir. The company attempts to buffer fluctuations from Brownlee Dam by releasing water a bit more steadily from Hells Canyon Dam. When possible, a minimum of at least 6,530 ft^3/s is released from Hells Canyon Dam to meet instream flow requirements for fish and navigation (Dennis Womack, Idaho Power Company, oral commun., 1993).

The dams of the Hells Canyon Complex were not designed specifically for flood control, and unexpectedly high flows certainly can occur on the Snake River. The consummate unexpected flow happened 15,000 years ago when Lake Bonneville (ancestor of today's Great Salt Lake) suddenly cut a new channel into the upper Snake basin. A peak flow of 20 to 33 million ft^3/s exploded through Hells Canyon, almost three times greater than the largest flood ever measured on the Amazon River (Jarrett and Malde, 1987, O'Connor, 1993). Historically, the more typical pre-dam floods would roll through Hells Canyon in May and June, reaching peaks of 75,000 to 95,000 ft^3/s every few years. But even with all three dams in place, floods of 74,000 ft^3/s or more have passed through the Complex on at least five occasions since 1970. Peak flow past the dams in a 1982 flood was 87,780 ft^3/s.

The great difference between the pre-dam and post-dam floods lies not in their peak flow, but with their sediment content. The three dams of the Hells Canyon Complex act as very effective sediment traps. Most suspended sediment reaching Brownlee Reservoir drops to the bottom of the lake; what little passes through is trapped behind the two reservoirs immediately downstream. Water released by Hells Canyon Dam is usually crystal clear. And no significant sediment-bearing rivers join the Snake until the Salmon River comes in, 60 miles downstream.

How much sediment are we talking about? The Snake River below its confluence with the Salmon carried as much as 5 million tons downstream each year before the dams were built (Jones and Seitz, 1980). The current status of the sediment balance in the system is unknown. One approach to quantifying sediment on the Snake River would be to sound the bottoms of the three Hells Canyon Complex reservoirs, especially Brownlee,

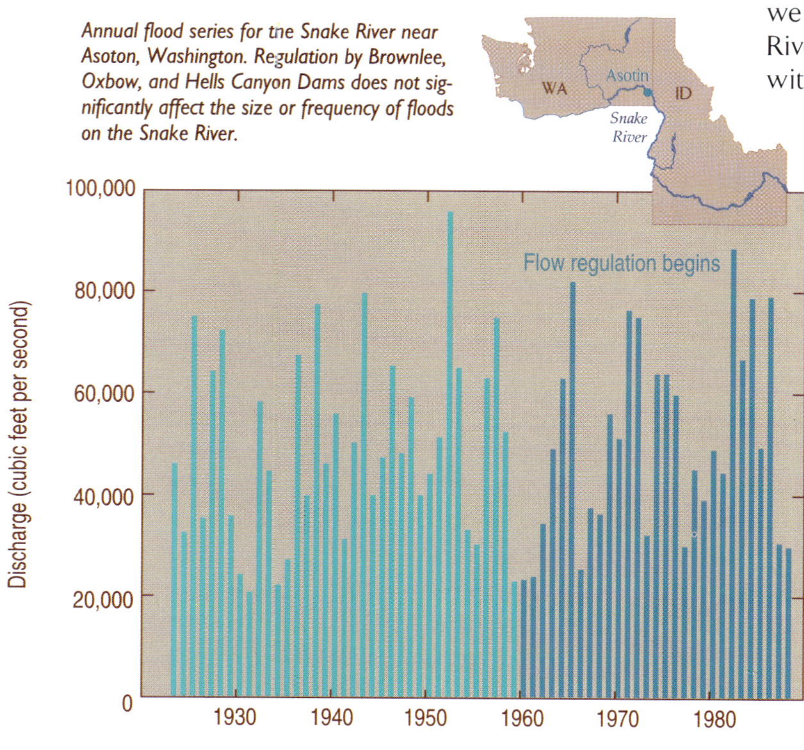

Annual flood series for the Snake River near Asoton, Washington. Regulation by Brownlee, Oxbow, and Hells Canyon Dams does not significantly affect the size or frequency of floods on the Snake River.

and compare the lake bottoms now to the shape of the pre-dam canyons. Such work has not yet been undertaken (Dennis Womack, Idaho Power Company, oral commun., 1993).

Or one could examine beaches along the Snake River below Hells Canyon Dam. Paul Grams (1991) did just that by comparing five sets of aerial photos of the canyon taken at intervals between 1955 and 1982. He found that the surface area of beaches in Hells Canyon had shrunk by 75 percent. Beaches between Hells Canyon Dam and the confluence of the Salmon were most heavily degraded; the Salmon appears to be reintroducing enough sediment to stabilize beaches below its confluence with the Snake. The greatest losses to beaches within Hells Canyon occurred from 1970 through 1973. But Brownlee Dam had been completed in 1958 and Hells Canyon Dam in 1967. Why didn't beach degradation begin immediately after closure? Grams concluded it had taken that long to flush enough sediment from the bed of the river before beach degradation could begin in earnest. As long as some critical amount of sediment remained in the system, floods continued to deposit as well as erode beaches. But once a river bed is stripped of sand, floods can take what little is left of the beaches and give nothing back.

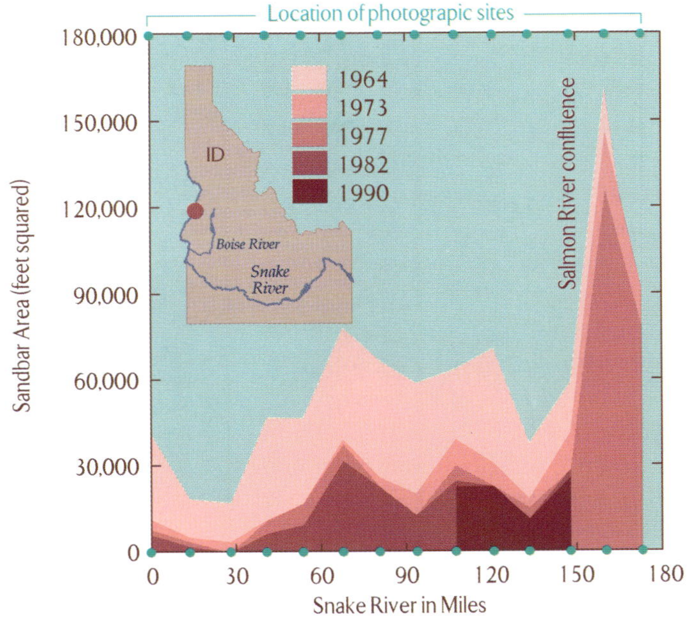

Sandbars on the Snake River in Hells Canyon have been shrinking since monitoring was begun in 1964; the greatest decrease is documented between photographs taken in 1964 and 1973 (modified from Grams, 1991; Schmidt and others, 1995).

Grams' study suggested that the beaches of Hells Canyon continue to shrink with each passing flood, but at a rate that has been decelerating since 1973. Does it matter? River runners think so; with each passing flood, they are more likely to be forced to camp in rocky sites amidst the poison ivy off the river, as beach after beach gradually disappears. How does one assign a value to the river runners' inconvenience? How does one compare the value of landscape relative to society's need for energy, or Idaho Power Company's responsibility to its customers and shareholders? Should something be done just for the sake of preserving the ecosystem of Hells Canyon? After all, the Snake carries the Wild and Scenic River designation and is within a National Recreation Area. But these areas were designated in 1975, long after Idaho Power Company was licensed to build the dams.

The Snake River below Hells Canyon Dam offers a classic example of degradation immediately downstream from a dam. Williams and Wolman (1984) documented many examples of such degradation below other dams. They concluded that the process begins with dam closure, occurs at a fast pace for the first few years, then slows with time until the bed is "armored" with material too coarse to be moved by the river, or until a bedrock control is exposed. This time-dependent process moves downstream as progressively more sediment is lost from the system.

Will Hells Canyon ever recover its beaches? Probably not in our lifetime. Alternative management scenarios for minimizing erosion exist, but they all have significant drawbacks. The least sophisticated alternative would be to simply dismantle the three dams. To be realistic, the dams in Hells Canyon are not among that handful of candidates for removal. Alternatively, even though the Hells Canyon reservoirs are relatively small, they could be managed primarily for flood control rather than for hydroelectric power. The reservoirs would be kept as close to empty as possible, and floods would be released as slow steady flow into Hells Canyon. The beaches might erode more slowly, but riparian vegetation would colonize the remaining fine-grained substrate. But this would cost Idaho Power Company (more precisely Idaho Power Company's customers) millions of dollars in lost revenues. Such a flow regime could engender its own set of downstream problems, such as the invasion of vegetation along the river banks.

Another option would be to somehow pass sediment through the dams. Idaho Power Company could devise a way to transport sediment around its turbines, thereby preserving the usefulness of its dams and simultaneously reintroducing sand to the sediment-starved reaches of the Snake River in

Sandbar on the Snake River in Hells Canyon

Hells Canyon. This alternative may seem attractive, but retrofitting this kind of technology would carry a frighteningly large price tag. And even if sediment could somehow be introduced to the river channel, that alone might not be enough to rescue the beaches or help the salmon. Once scrutinized, none of these alternatives offer a balanced solution to the issue of downstream effects at Hells Canyon. As much as we might wish to work miracles, sometimes the only fruit born of the study of the downstream effects of dams is a realistic assessment of relative values and environmental costs.

Rio Grande

Irrigation, Channel Aggradation

The Rio Grande is only accidentally a river between Elephant Butte Dam in New Mexico, and Presidio, Texas. The dam was designed to retain all flow on the Rio Grande, releasing water only for irrigation purposes. As a result, sediment has accumulated on the river bed and the channel has been invaded by tamarisk for hundreds of miles below the dam. Due to high runoff, unscheduled releases have occurred on a number of occasions since 1941. But the channel has diminished in size since the dam was constructed and can no longer contain floods as large as it once could. Consequently, relatively minor floods have caused significant damage downstream from the dam.

The Rio Grande drains more than 170,000 mi² of Mexico and the American Southwest (Bartlett, 1984), flowing 1,865 miles from the San Juan Mountains of Colorado to the Gulf of Mexico at Brownsville, Texas. In northern New Mexico, the Rio Grande is a vigorous snow-fed river, with a mean annual flow of 0.5 million acre-feet. Near the Gulf of Mexico, the river has an annual flow of 1.8 million acre-feet. But something strange happens in between. From Fort Quitman to Presidio, Texas, the river can be totally dry. At times, the Rio Grande is like one river cut in two, both ends alive but cleaved in the middle.

The story of the Rio Grande is one of withdrawal of water for agriculture and tributaries streams adding it back with sediment. Most of the water that flows through central New Mexico is redirected into canals in the El Paso area. But tributaries to this stretch of the Rio Grande are mostly unregulated, and runoff from periodic storms can

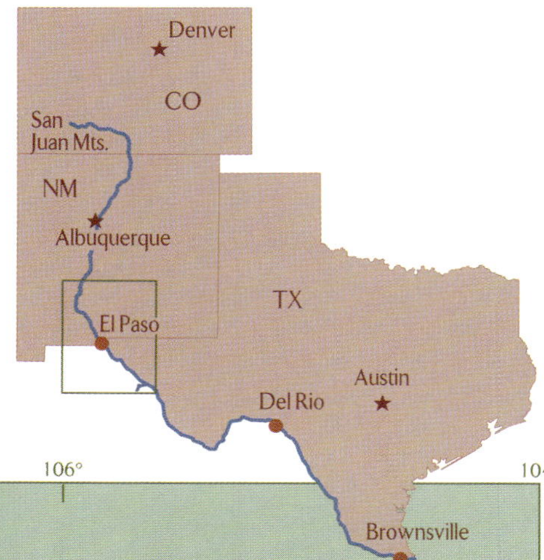

The Rio Grande between Elephant Butte Dam, New Mexico and Presidio, Texas

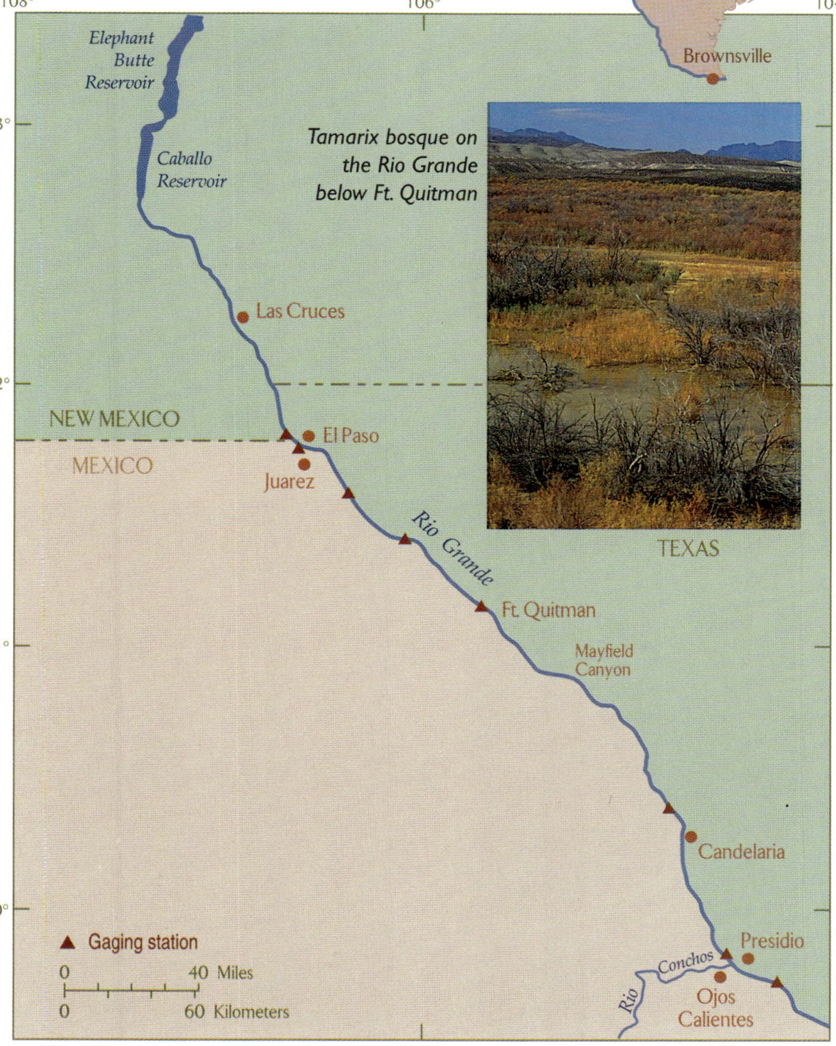

Tamarix bosque on the Rio Grande below Ft. Quitman

sporadically contribute both water and sediment. By the time the Rio Grande passes south-central Texas, it once again is a full-fledged river; large tributaries have restored much of the water diverted in New Mexico, west Texas, and Mexico. What happens in the middle illustrates how our attempts to control and use the flow of a river may backfire in a spiral of aggradation, canalization, and dredging.

The annual volume of flow in the Rio Grande above the Rio Conchos (at Presidio, Texas) has decreased dramatically with flow regulation. Gradual increases in irrigation of the Mesilla Valley of New Mexico have reduced the annual flow volume by an order of magnitude.

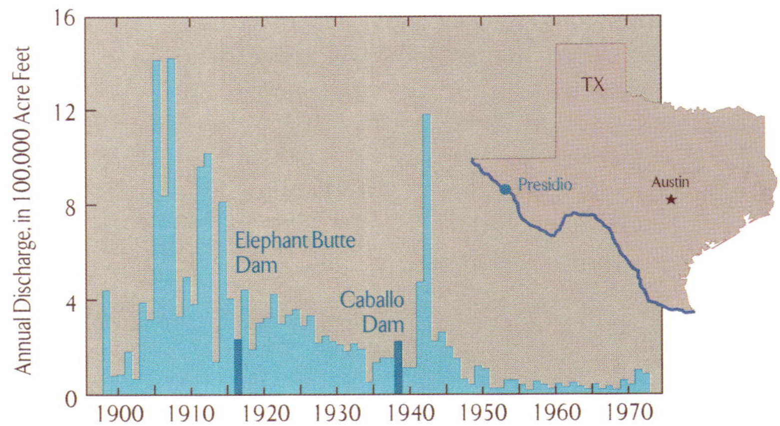

The Rio Grande in southern New Mexico and western Texas has more sediment than it can readily transport. The surrounding country is high Chihuahuan desert, held together by the roots of a few creosote bushes. When a storm passes through, tributaries of the Rio Grande don't just flow, they flood. For instance, Cibolo Creek joins the Rio Grande at Presidio. It is usually dry. But during a summer storm in 1990, it suddenly swelled to 6,000 ft^3/s (John Lee, International Boundary and Water Commission, oral commun., 1992). In 1904, a flood on Cibolo Creek washed most of Presidio away. These tributary floods transport tremendous quantities of sediment to the channel of the Rio Grande, producing alluvial fans that force the river against its opposite bank.

The Rio Grande's natural channel once maintained an uneasy equilibrium with incoming tributary sediment. The channel bed would aggrade next to tributary fans; flood plains would

The Rio Grande between Ft. Quitman and Presidio

grow above the river during moderate over-bank floods. Then a larger flood would pass, just powerful enough to carry away sediments that had accumulated at the mouths of tributaries. The flood would deepen and widen the channel, beveling down the tributaries' alluvial fans. Thus the channel form was preserved. But even in its natural state, the middle Rio Grande did not have the magnitude of flooding one might expect from a river that drains such an

Elephant Butte Dam, on the Rio Grande, New Mexico at Truth or Consequences

immense area. The largest flood on record at El Paso was only 24,000 ft^3/s, measured on June 12, 1905 (International Boundary and Water Commission, 1989).

Stream gaging began at El Paso, Texas in 1889, twenty-seven years before dams spanned the Rio Grande. Before Elephant Butte Dam was completed in 1916, the Rio Grande at El Paso had an average annual flow of 916,430 acre-feet, with most arriving during spring runoff in May and June. But some years only 100,000 acre-feet flowed by El Paso. A decade of predominantly dry years through 1904 prompted consideration of Elephant Butte Dam midway between Albuquerque, New Mexico and El Paso, Texas. Large-scale farming was to be possible in the fertile Mesilla and El Paso/ Juarez valleys only if the Rio Grande could be converted to a dependable year-round source of water for irrigation. During that turn-of-the-century

The Rio Grande below Ft. Quitman, Texas

drought, the Rio Grande had dried up too frequently to be trusted.

Construction of Elephant Butte Dam began in 1912. This was to be the centerpiece of the Rio Grande Project, an early offspring of the 1902 Reclamation Act that fathered the Bureau of Reclamation (U.S. Department of the Interior, 1981). Elephant Butte Dam's initial reservoir capacity was 2.6 million acre-feet, enough to store almost three average years of the river's flow. With the dam's completion in 1916, irrigation waters could be guaranteed to farms on two hundred thousand acres (U.S. Department of the Interior, 1981) that lay downstream in southern New Mexico and western Texas.

Riverside Dam on the Rio Grande downstream from El Paso, Texas

One consequence of this dependable water was that peak spring runoff averaging 4,400 ft^3/s in pre-dam years, was suddenly slashed to 1,300 ft^3/s (Mueller, 1975). The Rio Grande below Elephant Butte was transformed into a very different river: a river stripped of its floods. Below Elephant Butte Dam, the Rio Grande became a clear-water stream. Starved of sediment, the river scoured approximately two feet from the bed of its channel from Elephant Butte to Las Cruces during the first fifteen years of the dam's operation (Lagasse, 1980). This sediment made its way downstream toward El Paso, where the river gradient flattened and the sediment dropped out of transport.

With Elephant Butte Dam in place, New Mexico and Texas farmers were able to reliably divert millions of cubic feet of water every year. The 1906 Water Allocation Treaty assured Mexico that the United States would deliver 60,000 acre-feet of irrigation water annually at the International Dam between El Paso and Juarez (Mueller, 1975). Drainage ditches from both countries returned water from their irrigated fields back to the river – water that then carried high concentrations of salt out to the river and on downstream. The price of dependable water can be higher for some people than others, higher than some people wish to pay.

Further change came to the Rio Grande with a program of channel rectification undertaken by the International Boundary and Water Commission (IBWC). The IBWC had grown out of an 1884 treaty between the United States and Mexico, stipulating that the international boundary from El Paso to

Brownsville would be the deepest channel of the Rio Grande. Despite the treaty, the river refused to hold still. The deepest channel migrated back and forth across the flood plain, acting more like an out-of-control fire hose than a well demarcated line of significant political importance. The IBWC's answer to this erratic river behavior was to stabilize the channel with unnaturally straight levees on both the United States and Mexican sides of the river. Between El Paso and Fort Quitman, meanders were bypassed and the channel was tidied into a 65-foot slot between levees. The old river distance of 155 miles was shortened to 88 miles (Mueller, 1975). Similarly, the Bureau of Reclamation straightened and deepened the river's channel from Elephant Butte to El Paso.

Tributary floods continued, of course, and sediment was frequently delivered to the channel. This tributary sediment, combined with sediment stripped from the river's banks and bed closer to Elephant Butte Dam, began to accumulate downstream. The Rio Grande's bed through El Paso rose almost 13 feet between 1907 and 1933 (Reinhardt, 1937). The years 1941 and 1942 were anomalously wet throughout the Rio Grande's upper drainage. Two million, eight hundred thousand acre-feet of water flowed into Elephant Butte Reservoir in 1941 (U.S. Department of the Interior, 1981), and for the first time, water not earmarked for irrigation had to be released from the dam. The smaller Caballo Dam, built 22 miles downstream in 1938, could not dampen the flood. The river crested in El Paso on May 18, 1942, with a measured discharge of 7,700 ft^3/s. If the river's natural channel had not been clogged with sediment, this flood might have passed El Paso without problems. But on May 18, 1942, the lower reaches of town were afloat.

Despite efforts by the Bureau and IBWC, sediment still piles up in the Rio Grande faster than man or nature can remove it. Since 1951, the annual flow just upstream from Presidio is only 30,000 acre-feet (Everitt, 1993), one-tenth of the pre-dam average. The tributaries continued to add sediment, and the channel continued to aggrade. In 1987, a flood of less than 6,000 ft^3/s again wreaked havoc in El Paso.

The relatively small flood of 1987 carried tremendous amounts of sediment downstream. When the sand-choked flood reached the end of the channelized section below

The Rio Grande between Ft. Quitman and Presidio, Texas

Fort Quitman, it encountered a river without a defined channel. Water fanned over the old flood plain, and sediment settled out. In places, fields were awash for two years before drying enough to support the IBWC's dredging equipment.

Lester Ray Talley moved to the country below El Paso with his family in 1934. He was five years old and the Great Depression was in full swing. To get started at farming, his father sharecropped land just above Fort Quitman. Lester guesses that 100 families once farmed along both banks of the Rio Grande from Fort Quitman to Ojos Calientes in the 1950s, but most have left now. Lester looks downstream as he says, "The river's all choked up; ain't good for nothing."

The IBWC calls the 155 miles of the Rio Grande between Fort Quitman and Presidio its Boundary Preservation Project, defined entirely by what has *not* been done there: no canalization, no rectification, no sand-and-gravel mining, no bridges (Everitt, 1993). There is little but tamarisk stretching down the flood plain as far as the eye can see. Tamarisk was first reported along the Rio Grande near Las Cruces in 1910 (Robinson, 1965). Once-dominant galleries of

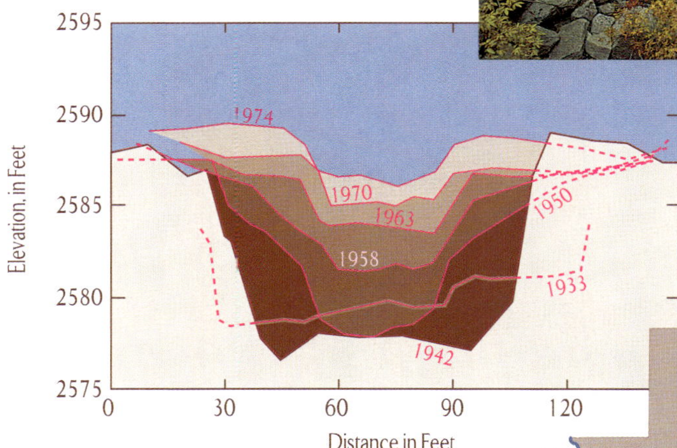

Time series of channel aggradation at Presidio, Texas, between 1933 and 1974. Because of the addition of sediment from tributaries, and the decreased flow in the Rio Grande, sediment is deposited in the channel. In response, the channel's bed elevation rises and its width narrows. (Modified from Everitt, 1993)

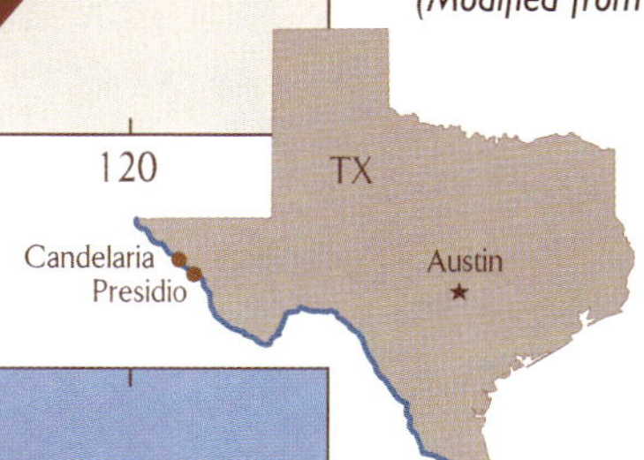

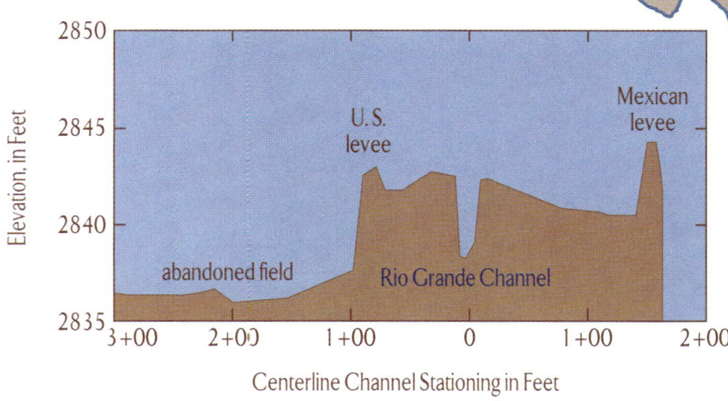

A cross section of the Rio Grande 2 km downstream from Candelaria, Texas. When measured in 1974, the channel bed was higher than its flood plain. (From Everitt, 1993)

cottonwood and willow along the Rio Grande have been overrun by this introduced species (Howe and Knopf, 1991). Tamarisk bosques in the Preservation Project do provide some understory for animal nesting, but when compared to the displaced native cottonwood/willow habitat, they offer little opportunity for feeding by birds and mammals accustomed to the area (Engel-Wilson and Ohmart, 1978).

Midway through the Preservation Project, Mayfield Canyon joins the Rio Grande in Box Canyon. Flash floods down this tributary have dumped sand, gravel, cobbles, and boulders onto the bed of the Rio Grande. Once floods on the larger river would have been able to clean out this debris. But regulation at Elephant

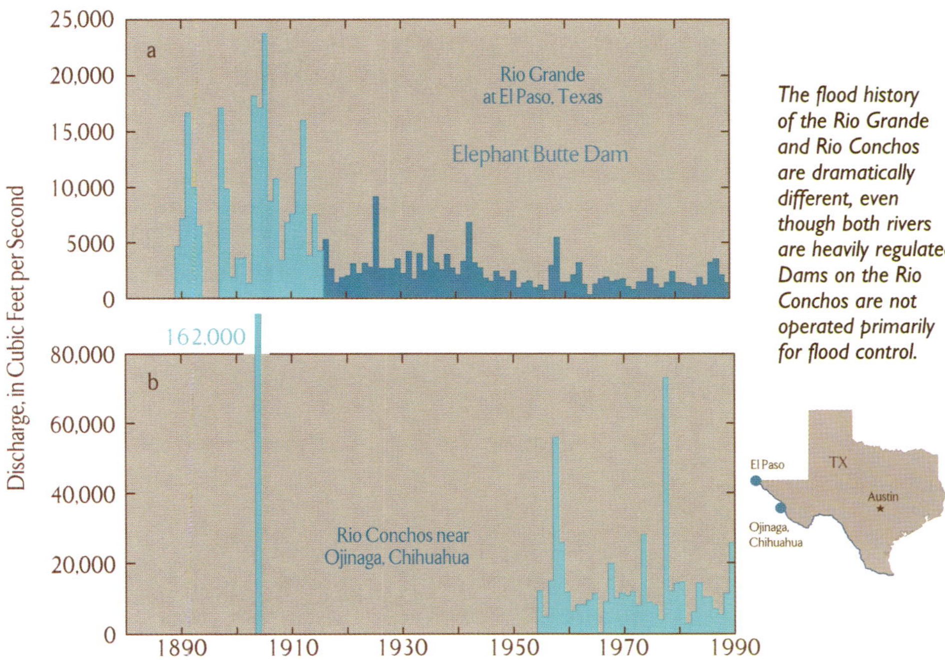

The flood history of the Rio Grande and Rio Conchos are dramatically different, even though both rivers are heavily regulated. Dams on the Rio Conchos are not operated primarily for flood control.

Butte, water withdrawals for irrigation and municipal use, and the dampening effect of tamarisk bosques have all worked to reduce the discharge and power of any flood coming down the Rio Grande. Sand and some of the gravel are moved downstream, but the cobbles and boulders remain in place. The river is no longer able to supply downstream force at the magnitude required to move these larger obstacles. A steep rapid has formed at the mouth of Mayfield Canyon. Confined by the canyon walls, the river is not free to migrate away from the tributary's debris. Waters backing up on the Rio Grande behind this debris fan had deposited ten feet of flood-plain sediment between 1945 and 1979 (Everitt, 1993).

The Rio Grande between Presidio and Redford, Texas

This scenario of aggradation continues for 155 miles downstream to Presidio, Texas. A dramatic change occurs there as the Rio Conchos enters from Mexico. The Conchos is not a trivial tributary; a flood in 1904 is estimated to have peaked at 162,000 ft^3/s. Its total annual flow averaged 737,000 acre-feet per year through the 1980s, more than five times the flow of the Rio Grande measured just above its confluence with the Conchos (IBWC, 1989). And unlike the Rio Grande, the Rio Conchos has an ongoing history of intermittent high flows 71,300 ft^3/s in 1978, 45,900 ft^3/s in 1991. The effects of these high flows are immediately obvious to anyone looking at the Rio Grande channel below its confluence with the Conchos. That channel is wide and well-defined. The tamarisks have been beaten back by the flood waters. Below the Conchos, in Big Bend National Park, the Rio Grande starts looking less like a swamp and more like a river again.

As Lester Ray Talley said, the Rio Grande from Fort Quitman to Presidio is just all choked up. By and large, people have abandoned the river here. Could the situation be different? The IBWC could spend a lot of money to extend its rectification down through the Preservation Project. Rectification would certainly make the IBWC's job of maintaining this section of the international border easier, at least for a while. Lester Ray Talley thinks it would eventually allow farming along the river valley again.

Flood control has been a long-standing reason for dams. In the days before dams and tamarisk and artificial channels, occasional floods rumbled down the Rio Grande and maintained a channel where sediment and seasonal flows were in equilibrium. Could Elephant Butte and Caballo dams be operated in a different manner that would minimize the aggravation of aggradation? It is conceivable that intermittent flushing flows released from Elephant Butte Dam could minimize the damage caused by uncontrolled flooding. But Las Cruces, El Paso, and Juarez have grown a great deal since the last unrestrained flood passed through in 1905. Who would want to pick up the tab for the damages from even a small intentional flood sent down the Rio Grande?

The Rio Grande from Elephant Butte Dam to Presidio is a fine example of a river now caught in a web spun with good intentions. Dams reduce peak flows. Irrigation canals siphon off annual discharge. Canalization temporarily helps handle floods, but tributary sedimentation cripples the effectiveness of the canals. We cross our fingers as we build levees, watch the river bed fill with sediment, and raise the levees a little higher. We build flood and sediment retention dams across tributaries and dredge out the river's channel. All the while, the growing populations of local cities push onto flood plains that ultimately cannot be made flood-proof.

Any one action will cause a reaction somewhere else. Does the complexity of this water system preclude any effective measures for dealing with its problems? No. Do we need to come up with more than temporary solutions to parts of the problem? Obviously. We need to be aware of all the tools available to us in addressing problems of water delivery, flooding, and sedimentation. It is conceivable that intentional larger-than-normal flow releases will one day be used as such a tool.

Chattahoochee River

Recreation, Population

The Chattahoochee River is managed with recreation as a high priority. Lake Lanier above Buford Dam is the most heavily visited, Federally managed lake in the country. The Chattahoochee Corridor from Buford downstream to Atlanta is a prize trout fishery. Metropolitan Atlanta has ready access to the Chattahoochee River National Recreation Area along the corridor. Although Buford Dam is managed largely for recreation, it is also used to meet peak-power demands of the Southeast power grid. The consequent patterns of water release affect the safety of recreational users and have resulted in erosion of the river banks downstream.

After the Civil War, John Taylor's great-grandfather settled along the Chattahoochee River 40 miles upstream from Atlanta to farm and raise a family. John grew up farming along the Chattahoochee, and remembers the big floods of winter and early spring before Buford Dam was finished in 1956. He would row across the swollen river, up into the trees where his neighbor Doug Milam now lives. The flood waters were thick with mud and sand, receding after a few days to reveal a new layer of organic-rich silt that lined the riverbank and spread across his fields. On January 8, 1946, 53,000 ft^3/s flowed through here (Stokes and others, 1991). These days, the river is dammed and discharge is rarely a fifth that of the old floods. John doesn't miss those floods, but he feels that the daily fluctuating releases from Buford Dam are ultimately more detrimental to his land. The new regime of clear water and daily fluctuation steadily erode a little more soil from the river's banks each year, but gives back nothing.

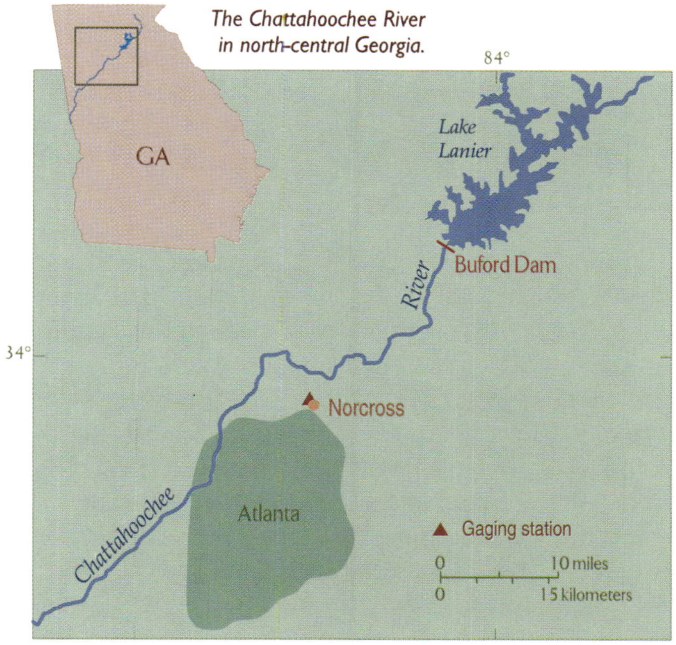

The Chattahoochee River in north-central Georgia.

The U.S. Army Corps of Engineers started construction of Buford Dam in 1952; four years and \$45 million later, the dam was completed and Lake Sidney Lanier began to fill. Buford Dam was intended to provide flood control, improved navigation, hydroelectric power, and a reliable supply of instream water as the river passed Atlanta. More dams have been built downstream on the Chattahoochee, and navigation is no longer considered feasible on this stretch of the river. The dam has admirably performed the tasks of flood control and instream water supply. Since completion, no destructive floods have occurred on the Chattahoochee below Buford. And the river is far and away the most important (and usually reliable) source of drinking water for millions of people downstream.

Fisherman on Lake Sidney Lanier above Buford Dam in Georgia

Flow records for the Chattahoochee River at Norcross, Georgia. Buford Dam has been successful in controlling floods, but the total annual flow has remained unchanged.

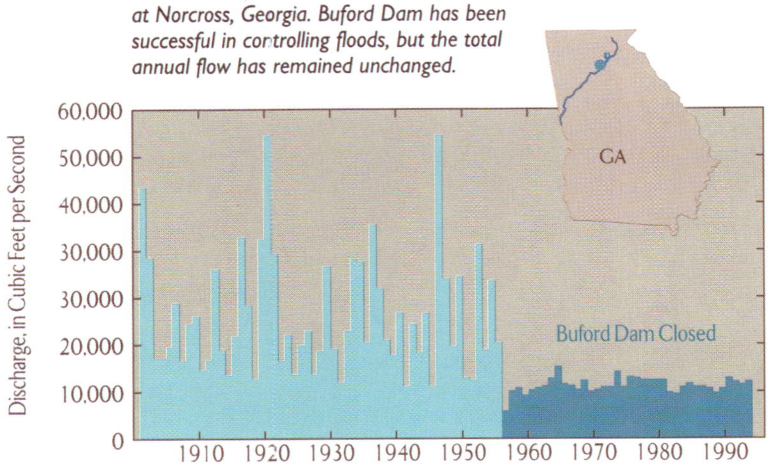

Greater metropolitan Atlanta is an energetic growth center for the New South: 2,500,000 people lived here in 1991 (Atlanta Regional Commission, 1992), with more than another million expected to arrive in the next two decades. By the year 2010, the Atlanta Regional Commission forecasts that the metropolitan area will require 200 acre-feet of water on an average day, and 310 acre-feet on a hot day (Atlanta Regional Commission, 1991). Virtually all of this water will come from the Chattahoochee River. The Corps is required to release enough water from Buford Dam to insure a minimum instream flow of 750 ft^3/s (1,490 acre-feet per day) as the Chattahoochee passes Atlanta. The city would like to see the flow increased, giving it more leeway in taking out water for its own uses (Atlanta Regional Commission, 1991).

A large and growing population consumes considerable electrical power. The Corps of Engineers has continued to abide by its mandate to generate electrical power at Buford Dam. The dam's three generators have a combined capacity of 86,000 kw. Typically, off-peak flow out of Buford is reduced to 1,300 ft^3/s on weekdays, and 600 ft^3/s on weekends, thus saving water for peak-time release (Benton Odom, U.S. Army Corps of Engineers, oral commun., 1992). A consortium of electric utility companies, along with the water users

Canoeing on the Chattahoochee River in Atlanta, Georgia

represented by the Atlanta Regional Commission, examine the Corps' weekly water allocation and assign an hour-by-hour schedule of releases for the upcoming week (Atlanta Regional Commission, 1991).

Flow releases in the Chattahoochee are timed to augment the Southeastern power grid's peak demands, when electrical power is most valuable. Eight-thousand, eight hundred ft^3/s are released during maximum power generation at Buford Dam, any flow higher would necessarily be sent through the spillway and wasted for power generation. As long as too much water isn't sent downstream over the course of the week, and as long as the river always measures at least 750 ft^3/s in Atlanta, the electrical companies are basically free to shape the release from Buford Dam any way they please (Michael Wilder, Georgia Power Company, oral commun., 1992).

A curious development has occurred along the Chattahoochee, however. The people of northern Georgia have fallen in love with their reservoir and its water. The Corps of Engineers has recognized

Eroded banks two miles below Buford Dam

The Chattahoochee River below Buford Dam

Warning on the Chattahoochee River immediately below Buford Dam

recreation as a priority by which to operate Buford Dam. In 1990, nineteen million people came to visit Lake Sidney Lanier, more than any other federally managed reservoir in the country. Twelve thousand homes surround this 38,000-acre reservoir. Twenty thousand boats are parked at 6,700 docks on the lake. The docks can tolerate some fluctuation of lake level, but not much. All those people spent $422 million recreating here in 1990 (Montgomery, 1991). By contrast, hydropower generation of electricity at Buford earned $1.3 million during the year ending September 30, 1992.

In 1978, Congress established the Chattahoochee National Recreation Area. It comprises fourteen units scattered along the river between Buford Dam and Atlanta. As land and funding become available for future purchases, the recreation area will expand to its authorized total of 6,800 acres. Predictably, the recreation area is heavily used by the millions of people who live within just a few minutes' drive. There they all are, furiously fishing, canoeing, bicycling, bird-watching, picnicking, jogging, and swimming. The National Park Service, charged with administration of the recreation area, worries about visitor safety (John Hendrix, Chattahoochee River National Recreation Area, oral commun., 1992). Does the river's water quality meet standards safe for swimming? Will canoeists be caught in the tangle of fallen trees that litter the channel? Will fishermen be surprised by the sudden daily rise of river level?

Signs warning about fluctuating dam releases have been erected every few hundred feet along the Chattahoochee's banks for many miles below Buford Dam. A series of ear-splitting sirens go off every time the river is about to rise. There are even low-powered AM radio stations broadcasting the changes of river level. Nonetheless, fishermen are swept downstream every year. Recreational safety is a major issue along this regulated river.

Trout do not naturally thrive in the typically warms waters of Georgia. But water from deep within Lake Sidney Lanier is cold and clear, ideal for trout. The State's Buford Hatchery releases hundreds of thousands of 10- to 12-inch trout every year to the waiting lures of folks lining the banks downstream (William Couch, Buford Fish Hatchery, oral commun., 1992).

Monitoring Water Quality in Regulated Rivers

A river's health can be assessed by many measures: the volume of water carried, the size of the channel, and the amount of sand on its banks. But rivers transport more than water. Organic and inorganic substances, either dissolved or attached to sed ments or organic debris, are all integral components of flowing water. Dams can dramatically increase or decrease transport of these substances. For example, salinity frequently increases downstream from dams on most western rivers because of the combination of reservoir evaporation and the addition of runoff from salt-rich strata. In the east, algae and nutrients can increase below dams, or dissolved oxygen can decrease. Decreased sediment load increases the amount of light passing through the water, further altering the biotic system.

Measurement of water quality in rivers requires more than simply filling a bottle and "testing" it. Water samples from two nearby points in a river may yield significantly different results. Hydrologists quantify sediment and nutrient loads, typically expressed in units of weight per day. Calculation of loads requires measurement of discharge at a stream-gaging station and periodic collection of samples that are discharge-weighted, or representative of the total flow through a cross section. Water quality can also be assessed by analyzing the biota; for example, fish concentrate metals and other contaminants in their body tissue. The types and density of organisms is another measure of water quality. Sampling of water quality is very expensive but is what is necessary to answer basic questions about the integrity of our nation's rivers.

The Game and Fish Department monitors water temperatures as well as the amount of iron, manganese, and dissolved oxygen in the Chattahoochee below Buford Dam. When water is drawn from near the bottom of the lake, oxygen concentrations can drop as low as half a part per million, well below the 5-10 ppm level considered desirable for trout. Recreation has become a driving force on the Chattahoochee. So have water quality, and fish and wildlife concerns. All of these priorities have taken their place alongside the four authorizations by which Buford Dam was originally operated.

In 1975, the U.S. Geological Survey undertook a three-year study of the hydrology and water quality of the upper Chattahoochee basin. The study looked in detail at the river's biochemical oxygen demand, dissolved oxygen, phosphates, thermal pollution, and other measures of water quality (Cherry and others, 1980). Engineers modeled the flow of water in order to predict its movement downstream from Buford (Jobson and Keefer, 1979). Strategies of dam operation were suggested that could most economically deal with problems of dissolved oxygen (Schefter and Hirsch, 1980). Geologists investigated the rate at which sediment was being swept into the river from nearby farms and logging operations (Faye and others, 1980). The State of Georgia made good use of the data that arose from the project, they legislated formation of the Atlanta Regional Commission, charged with design and implementation of a Chattahoochee Corridor Plan. This plan has been the vehicle by which a good deal of scientific thinking was converted into the nuts and bolts of municipal planning and zoning.

Two-hundred and nineteen thousand tons of sediment are annually carried into

the upper reaches of Lake Lanier (Montgomery, 1991). But at the other end of the lake, waters emerging from Buford Dam are clear and devoid of sediment. Below the dam, the Chattahoochee River is vigorously eroding its banks. John Taylor will tell you all about that, he feels that the river's daily fluctuations are hastening the erosion of his farm. If he had his way, the river would flow without fluctuation. If the thirsty city of Atlanta had its way, the river would have a higher minimum flow, thus necessitating lower peak flows. If fishermen had their way, releases from Buford would be cold water with adequate oxygen for their introduced trout. If boaters on Lake Lanier had their way, the lake would vary not an inch from its normal pool level of 1,070 feet above sea level.

So many demands but just one river. Dam releases could be designed to minimize downstream erosion, but power generation would suffer. Likewise, reduction of extreme fluctuations in flow would increase recreational safety but diminish the capability of load-following power generation. The operation of Buford Dam has been subject to a lot of tinkering since the dam's original authorization and construction in the 1950s. More changes are likely to come in the future. The inexorable growth of cities along the river will put water supply, water quality, and recreational interests at an even greater premium. The Army Corps of Engineers is likely to hold sway as far as flood control is concerned, but water allocations for generation of peak power could become an historical curiosity when that value is eclipsed by environmental and recreational needs that may, in time, become higher priorities.

Lake Sidney Lanier

Platte River

Channel Narrowing

Water developments have deple
much of the Platte River's volum
and significantly reduced the ma
nitude of spring floods. As a res
the channel has narrowed to as
little as 15 percent of its former
width. Sandhill cranes roost alo
the river during their spring migr
tion. But changes in the channel
morphology have increasingly
restricted the cranes' habitat.

Dusk settles over the Platte River as the sky fills with lesser sandhill cranes gliding in to roost. Their five-foot wings are extended but not beating, bony legs already pointing toward earth. The sprawling flock of birds looks like thousands of gray umbrellas, all open, all drifting down to earth. The air is thick with the cranes' chattering cry. In February and March every year, almost a half million sandhill cranes return to roost along the Platte River in a scene that has been repeated annually since the Ice Age (Krapu and others, 1982). The river has always offered habitat that the cranes need: shallow water spread across a wide channel, broken up by numerous sand spits and islands. But during this century, that riparian habitat has been drastically altered, largely by the placement of dams upstream.

Each morning in March before daybreak, the air above the river is filled with a rising chorus of cranes: a dry crackling cacophony that swells as the sun rises. An occasional bird lifts from the islands, and then everywhere flights of cranes launch skyward. Initially their movements are labored, uncertain. But after a few seconds the long ungainly legs are stowed aft and the great wingbeats become synchronous, almost stylized. The cranes are off for another day of foraging. The birds might fly a few miles before descending onto nearby cornfields that have been dormant since the previous fall's harvest. During its six or eight weeks along the Platte, each crane will add 15 percent or more to its wintertime body weight of seven pounds. Corn pecked from the farms offers more than 90 percent of this nutrition; the balance is gathered among meadows along the river where the birds find earthworms, snails, grasshoppers, and other delectables (Reinecke and Krapu, 1986). At day's end, the flocks once again return to their favored sandbars along the Platte.

The Platte is a quintessential river of the Great Plains: flat, wide, and shallow. Its banks were a major thoroughfare for the westward migration of the mid-1800s. Travelers consistently commented upon the river's sand-choked waters; they warned each other that no trees were available along its course for fuelwood. The river's main tributaries, the North and South Platte, reach like two great pincers around the Rocky Mountains of Colorado and Wyoming. Coming down out of the mountains, these tributaries encounter the relatively dry country of southeastern Wyoming, northeastern Colorado, and then western Nebraska. Rainfall along the North Platte as it flows from Wyoming into Nebraska is only 13 in/yr (Eschner and others, 1983). Beyond the shadow of the Rockies, precipitation increases to 25 in/yr at Grand Island, Nebraska (Williams, 1978) and even more as the river flows farther to the east. The upper Platte and its tributaries depend on springtime snowmelt from the Rockies to provide the greatest part of its annual flow.

The confluence of the North and South Platte Rivers lies just west of the hundredth meridian. John Wesley Powell, a famous western explorer and one of the founders of the U.S. Geological Survey, had observed the dramatic decline of rainfall across the Great Plains (Powell, 1879). Growing up on a Midwest farm, he knew how difficult the life of a farmer could be. And without reliable water, farming that might have just been difficult becomes impossible. All through the 1870s Powell championed the importance of irrigation as the prerequisite for organizing and settling agricultural lands west of the hundredth meridian.

The Platte River below North Platte, Nebraska

Primitive irrigation was practiced in the upper South Platte drainage as early as 1838, when Antoine Janis diverted water from the Cache la Poudre River onto his fields within the river's flood plain. Twenty years later, spurred by the demands of the 1858 gold rush in the mountains west of Denver, the pace of water diversion to irrigated farmland began to increase. Canals proliferated out from the river onto benchlands above the flood plain like cracks radiating through shattered glass. By 1885, more water had been appropriated by canal builders and farmers than actually flowed in the South Platte during the summer irrigating season (Eschner and others, 1983). In contrast, the North Platte was never the focus of a mining frenzy and developed more slowly. Cattlemen in Wyoming gradually embraced the benefits of irrigating fields to provide feed for their stock. It would be 1917 before the entire North Platte River was over-appropriated during summer.

In 1890, the Platte River Valley was well on its way to becoming the embodiment of Powell's ideal of irrigation and reclamation when the Western Frontier was officially closed. By

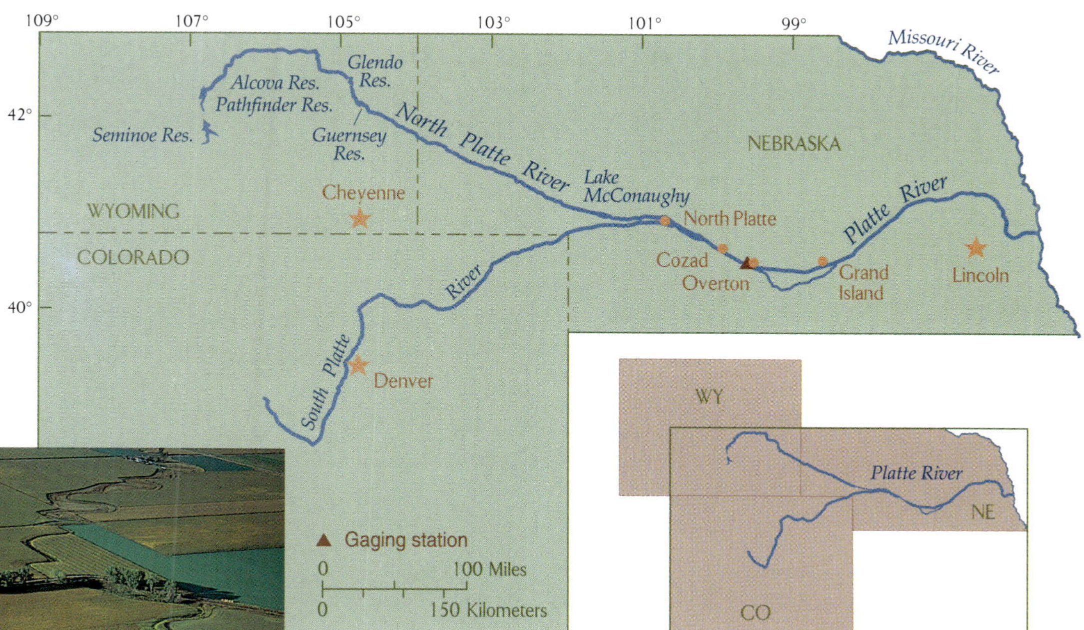

The Platte River in Wyoming, Colorado and Nebraska.

Farmland near Overton, Nebraska

1894 canals were irrigating 365,000 acres within the valley (Krapu and others, 1982). But by the turn of the century, canal construction began to decline as progressively less water could be wrung from the Platte at times when the water was most needed for irrigation. Private companies had built a few dams across small tributaries before the turn of the century, hoping to impound a bit of spring runoff for summertime use. But these were small structures and most of the Platte's springtime water still went downstream, slipping through the farmers' fingers. Who could afford

Keystone Dam on the Platte River

to build the larger dams desired by the farmers? The question was unequivocally answered by passage of the Federal Reclamation Act of 1902, which ushered in an era of big dams on the North Platte.

Flush with Federal money and armed with the best of intentions, dam builders descended upon the North Platte. Beginning with Pathfinder Dam in 1909, six major dams were built across this river in Wyoming and Nebraska. Seminoe, Pathfinder, Alcova, Glendo, and Guernsey reservoirs in Wyoming have a total storage capacity of 3 million acre-feet. Kingsley Dam was built in 1941, forming 1.9 million acre-foot McConaughy Reservoir just upstream from the river's confluence with the South Platte in central Nebraska (Williams, 1978). The South Platte, with fewer reservoir sites, was bridled with dams that could hold back only 1.3 million acre-feet (Krapu and others, 1982). Now water could be gathered year round and delivered when it was needed for irrigation in summer.

And delivered it was. With the reservoirs in place, demand for irrigation water expanded to meet the supply. Another wave of canal building swept down the North Platte,

Annual flow volumes for the Platte River at Overton, Nebraska. Five dams built since 1909 provide flow regulation for irrigation of farmland in Wyoming, Nebraska, and eastern Colorado.

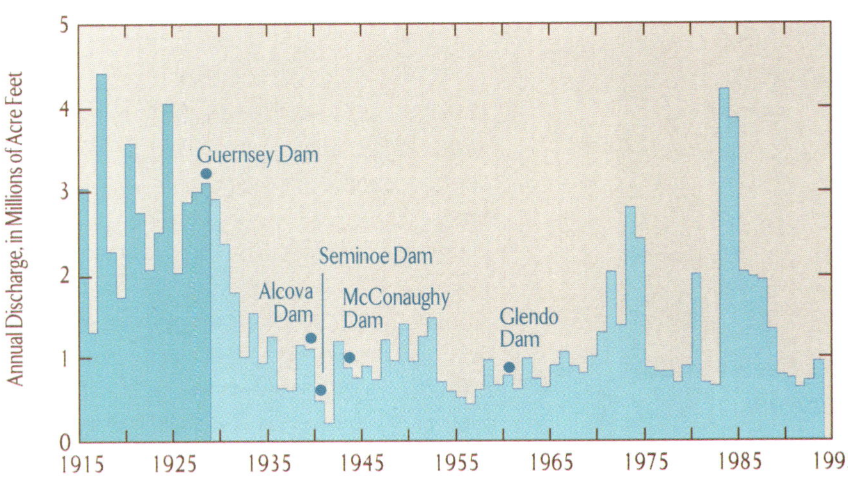

lasting until the 1930s. Even more water was sucked from the river, and farmers turned to ground water for additional irrigation. By 1979, only two other states had more land under irrigation than Nebraska. Annual flow of the Platte at Overton, Nebraska once averaged 2.9 million acre-feet per year; by 1984, less than one-third as much water would pass Overton during a typical year (Krapu and Eldridge, 1984). Peak springtime flows that had once rolled past the town of North Platte steadily dropped from a pre-dam average of 18,000 ft^3/s to 2,500 ft^3/s for the period 1957-1970 (Williams, 1978). Spring flooding had become a thing of the past.

Canadian and Snow Geese near the Platte River at Kearney, Nebraska

Still, the sandhill cranes keep returning. They congregate in greatest numbers from Overton to Grand Island along the Platte River. A smaller group is found on the North Platte River just above its confluence with the South Platte. The cranes roost on sandbars where the channel is at least 500-feet wide, and are rarely found in areas where the channel is less than 150-feet wide (Krapu and others, 1984). These preferences were not a problem one hundred years ago; a railroad survey in 1866 reported that the island-studded channel was 4,000- to 6,500-feet wide from North Platte down to Kearney. But the steady reduction of both peak springtime flows and total annual flows have taken their toll. In the absence of floods, cottonwood, elm, and willow have successfully invaded the bare sandbars. Islands that once isolated the cranes from their predators are now connected to the river's banks; sediments fill the intervening channels, and spring floods are no longer available to flush these minor channels clear (Eschner and others, 1983).

Sandhill cranes dancing near Kearney, Nebraska

Each spring, the cranes returning to the Platte find many stretches of the river a little bit narrower. In its 60-mile stretch above Overton, the channel by 1965 was only 10 to 20 percent of the width measured in 1865 (Williams, 1978). The cranes have abandoned a bit more river every year; by 1956, they had retreated from the 30-mile stretch from North Platte to Cozad. Each spring, the birds are packed a little bit tighter against each other. Crowding increases the likelihood of infectious disease. An avian cholera outbreak decimated tens of thousands of waterfowl in the Rainwater Basin, just a few miles from the cranes' roosting areas along the Platte (Krapu and others, 1982).

Unlike whooping cranes, sandhill cranes are not endangered. Despite the habitat reduction along the Platte, their numbers may actually be increasing. It is refreshing, for a change, to contemplate measures to insure the persistence of a species, not when it is tipped beyond the brink of extinction, but while it is still in its prime. Botanists have suggested that moderate releases of water in late spring could submerge sandbars that would otherwise host the germination of new cottonwoods and willows. And hydrologists now contemplate dam releases capable of opening and maintaining a channel adequate for use by the cranes. There is no cookbook method for determining the amount of water required to produce both of these results. A feedback loop of altered dam releases and subsequent downstream observation (known as Adaptive Management) would be required to determine how much water would have to be released for how long.

Part of the equation is cost: reduced irrigation water for farmers living upstream. The potential trade of farm productivity for crane habitat is not as simple as it might seem. The timing of releases might coincide with periods when farmers do not require water. Or crops could be converted to ones that require less irrigation. Although scientists can offer answers to many of these questions, the central question the relative value of cranes versus farm productivity can be answered only by our society.

Green River

A Biological Mandate

Construction and operation of Flaming Gorge Dam has altered channel morphology and water temperature in the Green River. Native fish were sufficiently threatened by these changes that the U.S. Fish and Wildlife Service invoked the Endangered Species Act to mandate dam releases deemed least likely to harm native fish. Studies are now underway that attempt to link the creation and maintenance of habitat to sedimentary processes influenced by dam operation.

Landscapes within the drainage of the Green River are reminiscent of more Western movies than you might ever want to watch. The 45,000 mi^2 area of Wyoming, Colorado, and Utah that contributes runoff to this river is a spectacular mix of mountains and high desert. The river's source in Wyoming's Wind River Range is 730 miles upstream from its confluence with the Colorado River in Utah's Canyonlands National Park. Along the way, the Green passes through Dinosaur National Monument

Canyonlands National Park

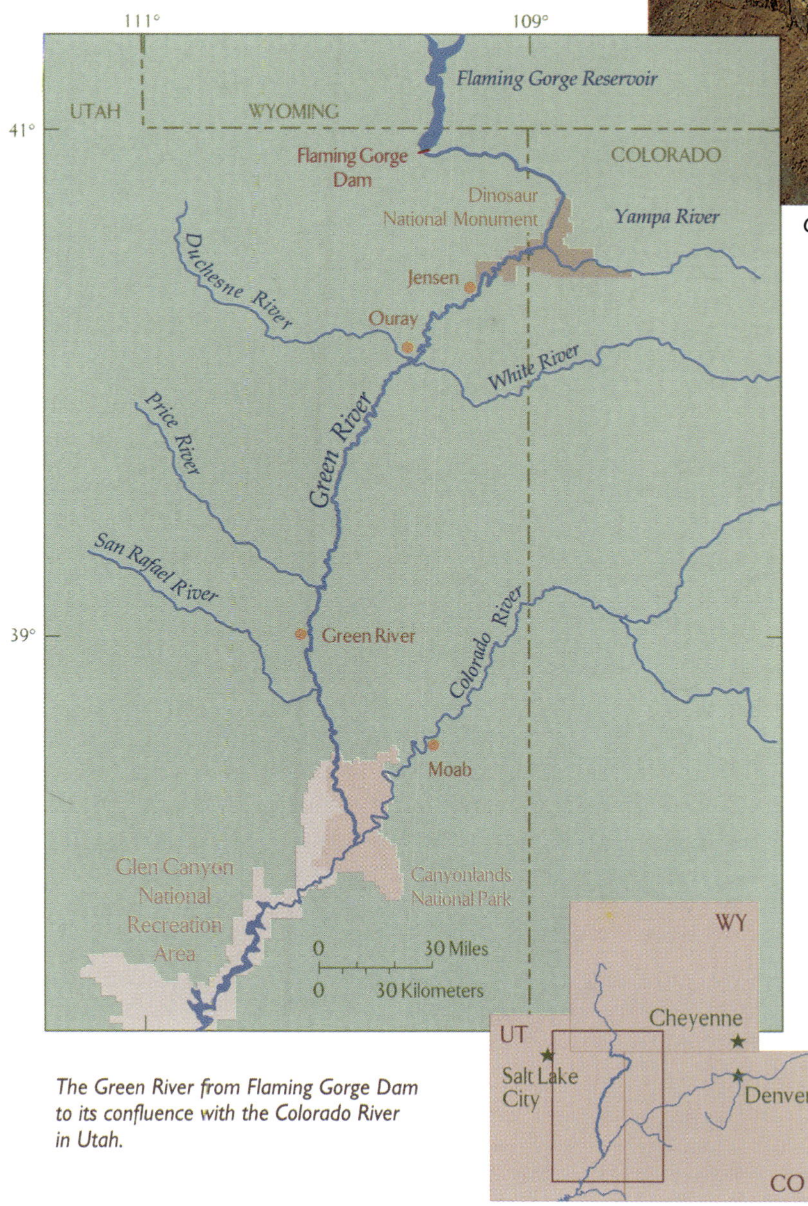

The Green River from Flaming Gorge Dam to its confluence with the Colorado River in Utah.

as well as several national recreation areas and wildlife refuges. It gathers the flow of the Yampa and White Rivers, which drain the Rocky Mountains of northwestern Colorado, as well as the Duchesne, Price, and San Rafael Rivers, which drain the Uinta Mountains, Wasatch Plateau, and the lower deserts of Utah. The high mountains and plateaus that form the rim of the Green River basin contribute most of the river's annual flow, even though they are only a small part of the entire drainage basin (Iorns and others, 1965).

Before construction of dams in the basin, the Green would begin to rise in March, swollen with snowmelt from its headwaters. The spring flood would peak in early June and not recede until July. The average peak flood at the most downstream gaging station — Green River at Green River, Utah — was 32,000 ft^3/s for the period from 1895 to 1962; and the highest peak ever measured at this station was 68,100 ft^3/s in 1917. Torrential spring floods would carry great loads of sand and silt, but only small amounts of sediment were derived from the water-producing headwater stretches of the river.

The Green River has a split personality: its water originates in the mountainous highlands, but its sediment is derived from the low-lying deserts. The forested headwaters yield little sediment during the spring snowmelt. The desert areas are subject to brief, violent summer thunderstorms, which strip sediment from the sparsely vegetated slopes. Prior to construction of Flaming Gorge Dam, most of this sediment was deposited on the channel bottom until spring floods could transport it downstream.

The Green River near Sand Wash, Utah

The Green River, like most other large rivers on the Colorado Plateau, once displayed a wide range of hydrologic behaviors. The water was nearly clear during low-flow periods. Cold spring flood waters were thickened with enough sediment to "stand up a spoon in a coffee cup." Sluggish summer flows were occasionally stirred into a muddy frenzy by the input of a local thunderstorm. Winter flows became cold enough above Jensen that the top two or three feet of the river would freeze solid.

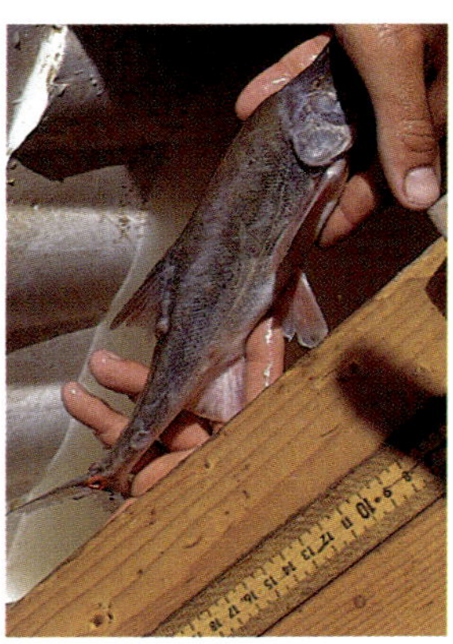

Humpback Chub

The extreme variability in flow, sediment concentration, and temperature of the Green River gave rise to an array of fish species found nowhere else in the world. The river was home to at least thirteen endemic species in the minnow, sucker, trout, and sculpin groups. This combination evolved into an interlocked community, balanced delicately amongst themselves and attuned to the demands of the river (Tyus and Karp, 1991). Seventy-four percent of fish first found in the Green and Colorado Rivers lived only in these rivers (Miller, 1959). Among these species are the razorback sucker *(Xyrauchen texanus)*, bonytail chub *(Gila elegans)*, humpback chub *(Gila cypha)*, and Colorado squawfish *(Ptychocheilus lucius)*. But today the bonytail chub may already be extinct, and the razorback sucker is only occasionally found. Each species developed unique adaptations to the rigors of life in the Green River and its own preference for different reaches of the river. The suckers, chubs, squawfish, and dace are now threatened by changes that have swept down the river since the 1960s.

For two million years the humpback chub has been coping with these vagaries, preferentially seeking the reaches in desert canyons with rapids. The chub is an odd-looking fish, with a pointy head and fins that seem disproportionately large compared to the rest of its body. The hump on its back is an adaptation that provides stabilization in swift currents. The humpback chub is not a particularly strong swimmer, instead it uses its

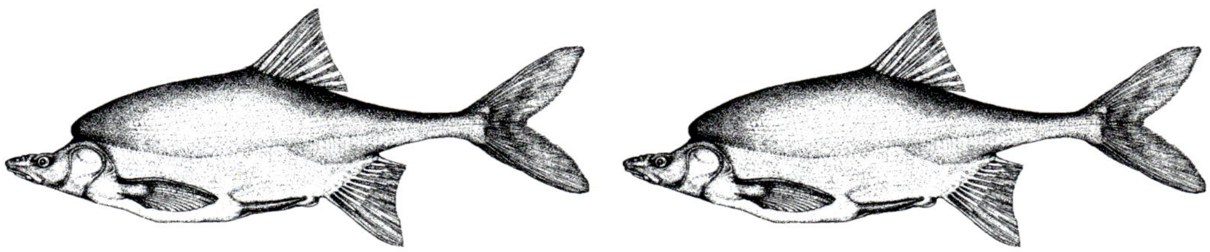

oversize fins to hydrodynamically "soar" underwater. Its lateral stripe is so sensitive that the chub can feel tiny vibrations caused by nearby insects — an ability well suited to life in muddy water with low visibility (Valdez, 1993). The humpback chub commands respect because it is a unique survivor. It demands concern because it is now endangered.

Native fishes can be threatened by many environmental factors, only some of which are directly related to dam construction. Many new species have been introduced to the Green River, among them channel catfish *(Ictalurus punctatus)*, red shiners *(Notropis lutrensis)*, and carp *(Cyprinus carpio)*. Collectively these species now greatly outnumber the native species. Some of these non-natives compete for the same food and habitats as the natives, and sometimes the natives (especially when young) are eaten by the non-natives.

The Green River entering Split Mountain in Dinosaur National Monument, Utah

Some dam-related stresses to the native fishes were inevitable, while others did not have to happen. Just before the gates of Flaming Gorge Dam closed in September 1962, a triumvirate of Federal and State agencies dumped 21,500 gallons of 5-percent rotenone into the Green River at various stations in Wyoming. The intent was to kill the "rough" fish that might interfere with stocked trout. Potassium permanganate was added to the brew 29 miles below the dam, in hopes of neutralizing the rotenone before it reached Dinosaur National Monument. The effect of this experiment was to kill significant numbers of the native fish that we now labor to save under the Endangered Species Act.

Flaming Gorge Dam has also been responsible for other direct and less avoidable effects on the native fish community. Because water is released from deep within the reservoir, water temperatures plummeted from a pre-dam average of 66°F to 42°F (Vanicek and Kramer, 1969). Although native fish can live in such cold waters, as they did each winter in the northern parts of the watershed, the low temperatures inhibit their spawning. For example, the Colorado squawfish is reluctant to spawn in water colder than 59°F; its optimum spawning conditions are closer to 71-77°F (Tyus, 1990). The cold water forced the squawfish (as well as most other native fish) to abandon the Green River from the dam down to the mouth of the warmer Yampa River (Vanicek and others, 1970).

The clear cold water of the regulated Green River became the home of introduced trout; the fishery has an international reputation which attracts millions of dollars in sport fishing each year. However, the water released from the dam was initially even too cold for trout – the low temperature inhibited their growth. In 1978, the Bureau of Reclamation spent $20 million to retrofit multilevel intake structures at Flaming Gorge Dam to allow for seasonal warming of the downstream waters to increase the growth rate of trout. This engineering solution enabled the Bureau to release waters from different levels and different temperatures. The trout grew fatter.

Flaming Gorge Reservoir filled for the first time in 1966; since then, the total amount of water released downstream has not significantly differed from the annual total prior to dam closure. However, the timing of water flow has been radically altered. With the dam in place, winter flows have increased and the spring flood has almost entirely disappeared. Flow almost never exceeds powerplant capacity.

The dam's powerplant is designed to generate 36,000 kw when 4,700 ft^3/s is released through the turbines at maximum reservoir capacity. Because peak electrical power demands occur during summer and winter, these are now the times of maximum river flow. Since the peak demands only exist for a few hours each day, the river would fluctuate so as to produce maximum power during the few hours each day when the needs are greatest. Immediately downstream from the dam, vertical fluctuations in daily river levels could be measured in yards. River levels 107 miles downstream at Jensen, Utah rise and fall as much as 1.5 feet each day (Valdez, 1989). These fluctuations can break up the ice cover, which is now thinner; the fish are forced to scatter as the ice grinds against the shore. Fluctuations in flow affect aquatic habitat and can compromise the ability of fish to survive through the winter.

Flaming Gorge Reservoir effectively intercepts all sediment that enters its headwaters. Water released from the dam is devoid of any sand or silt. Between the dam and its confluence with the Yampa, the clear water of the Green River is carrying away sediment from its banks and bed, and tributaries only replace some of it. This reach of the channel has been actively scoured in response to the dam. The situation changes when the Yampa River joins the Green 68 miles downstream from the dam. The Yampa has no dams along its entire course and is essentially unregulated. It plays a pivotal role in the viability of the Green River below Flaming Gorge Dam. With its high spring flows, sediment load, and warm summer waters, the Yampa is a sanctuary for the squawfish, chubs, and razorback suckers that continue to spawn in its waters.

The Yampa delivers most of the annual load of 3.2 million tons of sediment that the Green carries past Jensen, Utah since Flaming Gorge Dam was completed (Andrews,

1986). Throughout the 98 miles below Jensen, the Green's sediment budget is balanced – aggrading and degrading in approximately equal volumes since construction of the dam. Farther downstream, desert sidestreams contribute relatively more sediment than water to the Green. Now deprived of the large spring floods that once could move this sediment, the channel of the Green has aggraded and narrowed below Jensen, Utah. This aggradation has caused narrowing of the channel, infilling of secondary channels, and attachment of mid-channel bars to the river banks (Andrews, 1986).

The Green River above Sand Wash, Utah

Three and a half million acre-feet of water flow down the Green River every year. And every year, withdrawals totalling 1.3 million acre-feet are allotted for agriculture, mining, power plants, and consumptive use. In the late 1970s, the Central Utah Project began to take an additional 143,000 acre-feet from the Duchesne River through its Strawberry Aqueduct. But on February 27, 1980, the U.S. Fish and Wildlife Service invoked the Endangered Species Act, and issued a Biological Opinion in defense of the Green River's native fish population. The Service said that if more water was to be withdrawn at the Strawberry Aqueduct, then a reasonable and prudent alternative must be formulated to protect the fish elsewhere. The alternative all parties finally agreed upon was to modify the operation of Flaming Gorge Dam. No longer would dam releases be dictated by the timetables of irrigation, flood control, and peak-power demands. Instead, the dam would be operated in a manner that would protect the habitat and livelihood of the endangered chubs, suckers, and squawfish.

Biologists suddenly wielded a new and unusual tool with which to protect native fish: a 502-foot-high dam. Interim flows were instituted at Flaming Gorge Dam in 1985. Studies of fish carried out during the 1980s prompted refinements of the Biological Opinion. Dam releases are now seasonally adjusted to roughly mimic the river's pre-dam hydrograph. A steady 4,000 to 4,700 ft^3/s is released for one to four weeks in May to match the peak of the Yampa's spring flood. Flows are gradually reduced after this peak, until summer flows of 1,100 to 1,800 ft^3/s (measured below the confluence with the Yampa) are reached and then maintained throughout the summer. Small daily fluctuations are allowed during the summer, but are discouraged during wintertime.

The Green River through Canyonlands National Park

Estimating Sediment Budgets

Assessing the sediment budget downstream from a dam is a technique useful for developing a management plan for a regulated river. Calculating a sediment budget is similar to balancing a check book: one attempts to account for all sediment entering, leaving, and stored in a river system. The budget can be formulated as $S_i + S_t - S_o = \Delta S_r$, where S_o = the rate of sediment leaving the reach, S_i = the rate of sediment entering the reach, S_t = the rate of tributary additions, and ΔS_r = the rate of change in storage along the river's banks and bed. For most rivers downstream from dams, $S_i \approx 0$ because the reservoir traps most of the sediment that previously was transported downstream. S_o is measured at the gaging station at the downstream end of the reach and can be manipulated by changing the pattern of flow releases from the dam. ΔS_r, which also can be altered by flow releases, is measured using bed-monitoring techniques or remote sensing. S_t, which is delivered by small tributaries, can be measured or estimated. Tributary additions of sediment, which are affected by local geology, land use, and climate, are less easily altered to affect a regulated river.

A sediment budget can help explain the consequences of historical dam releases and flow regulation. Because the Salt River is unregulated, and Si is much greater than zero, ΔS_r is approximately zero over periods of decades. On rivers such as the Green below Flaming Gorge Dam, the Chattahoochee, and the Snake in Hells Canyon, where tributaries are small, S_t is much less than S_o. As a result, ΔS_r is negative and the rivers are scouring their beds and banks. For the Rio Grande and the Platte, S_t is much larger than S_o, resulting in a large, positive ΔS_r and aggradation of the channels. Tributary additions are a recognized problem along the Rio Grande, and sediment-retention structures span some tributaries in an attempt to reduce S_t. The Colorado River in Grand Canyon is a mixture of both types: large tributaries supply some sediment for S_t, and the pattern of flow releases from Glen Canyon Dam has reduced S_o. Because of this flexibility, releases from the dam can be modified to manipulate ΔS_r and redistribute sediment along the channel. In all cases, the sediment-budget approach is used to suggest the source of the problem and point to potential management solutions.

The redistribution of Flaming Gorge Dam water releases revolved around the survival of endangered native fish. Spring peaks are meant to facilitate spawning and protect young fish in backwaters. These high flows scour the channel bed and expose gravel that is necessary for egg attachment. Gradually declining peak flows are thought to help prepare backwater nursery habitats as well as stimulate specific biologic responses in native fish. During summertime, water of a temperature desirable to native fishes is released by using the dam's multilevel intake structures that were initially designed to aid introduced trout. By minimizing winter fluctuating flows, fish may be protected from unnecessary jostlings of the icepack beneath which they live. In short, dam releases throughout the year are used to promote native fish habitat as much as possible.

The operation of Flaming Gorge Dam has been shaped by engineers and scientists who wished to provide the benefits of power as well as an intact downstream ecosystem. Scientists from the Bureau of Reclamation were able to correlate discharge with backwater-habitat availability (Pucherelli and Clark, 1990). Fisheries scientists determined the life history, spawning cycle, and nursery habitats of native fish (Grabowski and Hiebert, 1989; Tyus and Karp, 1991; Valdez, 1993). USGS hydrologist Ned Andrews computed sediment budgets for successive segments of the river (Andrews, 1986) and, in cooperation with hydrologist Jon Nelson of the U.S. Geological Survey, developed a model of how the Green River adjusts its shape to high discharges (Andrews and Nelson, 1989). Ongoing research by the Bureau of Reclamation, Fish and Wildlife Service, States of Colorado and Utah, and Utah State University's College of Natural Resources is refining the interactive relations among reservoir releases, channel morphology, aquatic habitat formation and availability, and survival of endangered fish.

Time will tell if native fish can make a comeback on the Green River. Dam operations have been modified in ways thought to be beneficial to the habitat of native fish. But other factors, such as the presence of introduced species, also affect the fishes' chances for survival. The work done collaboratively by geologists, hydrologists, engineers, biologists, and economists on operations of Flaming Gorge Dam sets a precedent for a cooperative approach to minimizing the problems that exist below dams. Their methods and mathematical modeling can be applied to many rivers beyond the Green.

The Green River in Desolation Canyon, Utah

Colorado River
Multiple Mandates, Multiple Use

Glen Canyon Dam has altered sediment transport and biological communities along the Colorado River in Grand Canyon National Park. The authorization for Glen Canyon Dam is not only for water storage and distribution, but also for flood control, recreation, and hydroelectric power. Studies that provide scientific input into dam operation suggest the benefit of occasional high releases ("beach-building flows"). The cost of such a release can be contrasted with its benefits.

About five million people peer over the rims of Grand Canyon every year, and if they find themselves at the right spot, they may even see the Colorado River, almost a mile below. Some visitors will amble down the canyon's trails in the morning and struggle back to the rim later in the day. A few will stay overnight or longer, collectively spending a total of 80,000 nights a year camping in the canyon. Another 22,000 people annually spend six to as many as twenty-eight days each, rafting down the Colorado River, camping on its beaches. By any measure, Grand Canyon is one of the nation's most popular national parks.

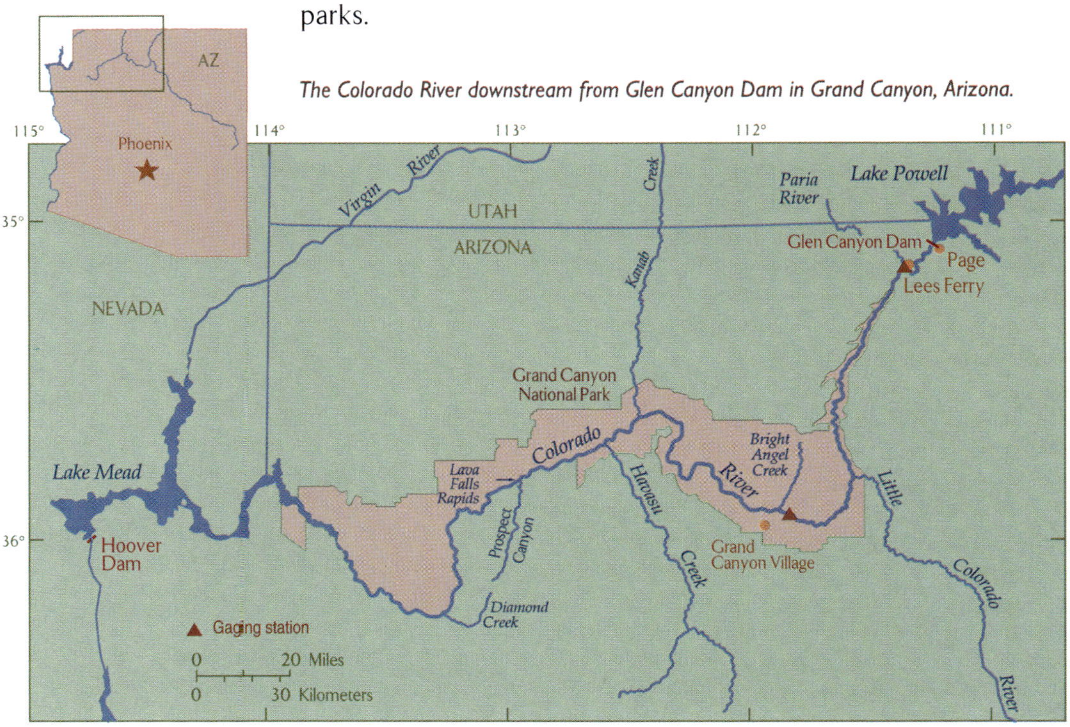

The Colorado River downstream from Glen Canyon Dam in Grand Canyon, Arizona.

Ironically, the river experience, though found within the deepest part of the canyon, is a trip through that part of the Grand Canyon environment most directly affected by human activity. That activity – the construction and daily operation of Glen Canyon Dam – is located 15 miles upstream from Lees Ferry, the launch point for all boating trips through Grand Canyon.

The Colorado River is the very heart of Grand Canyon. John Wesley Powell was first to float through the canyon in 1869. Every few miles, a major rapid punctuates the river's otherwise quiet course. These legendary rapids have colored the vision that we have of ourselves in the Great Outdoors: tiny heroic figures clinging to wooden boats, tossed and tipped by fifteen-foot waves. But the rapids constitute only a small portion of the river's overall length. The Colorado River has a great deal more to offer than simple amusement-park potential.

For the most part, the Colorado glides silent as a snake through the Canyon. Bits of river-litter – leaves and twigs, an occasional piece of tree root – dance along the squiggly line that separates the main current from an eddy flowing back upstream. Like once forgotten memories, vortices suddenly swirl to the surface without warning, created where the current encounters unseen obstructions that lie along the channel bed. Late in the day, sunstruck cliffs three thousand feet above the river cast their glow on the water's quiet surface: golden ripples, liquid reflections. The river seems at peace with the walls that confine its course.

Historically (*i.e.,* before 1963), an average of 12 million acre-feet of water rolled through Grand Canyon each year. The Colorado's flow followed a distinct seasonal pattern. The floods of May and June typically peaked at greater than 86,000 ft^3/s. When spring passed, the river dropped. Thunderstorms later in the summer would intermittently swell the river to perhaps 30,000 ft^3/s. By winter the river was usually reduced to little more than a small clear stream; January mean flows were only 5,260 ft^3/s before the dam was built.

Surveyor on the Colorado River in the Grand Canyon

A remarkable flood passed Lees Ferry on July 7, 1884. Warren Johnson was the ferryman; Jerry Johnson was his observant son. The flood was high enough to drive Jerry's cat well up into an apple tree. Johnson's recollection of that cat thirty-seven years later led E.C. LaRue to back-calculate the discharge of the 1884 flood at 300,000 ft^3/s (Howard and Dolan, 1981). It is likely that even larger floods have come through Grand Canyon within the last couple millennia. Flood deposits are perched well above normal river stage just downstream from Lees Ferry; they offer tantalizing evidence of a flood larger than 500,000 ft^3/s within the last 1,600 years (O'Connor and others, 1994).

In contrast to Flaming Gorge Dam, which is located upstream from the major sediment-producing tributaries of the Green River, Glen Canyon

The Colorado River at Lees Ferry, Arizona

Dam traps the vast quantity of sediment (66 million tons per year) that once flowed through Grand Canyon. The drainage area upstream from Lees Ferry is 111,800 mi². The water running off that area flows into Lake Powell from the high country of Wyoming, Colorado, and Utah, but the sediment has largely come from the desert country of southern Utah, southwest Colorado, and northwestern New Mexico. Desert tributaries in this region bear only 15 percent of the Colorado River's flow, but contribute 69 percent of its sediment.

Thus, the release of clear water into a canyon that once carried extremely high sediment loads is a recipe for substantial environmental change. Grand Canyon's sediment balance would be even more skewed if it were not for two downstream tributaries. Before construction of Glen Canyon Dam, the small Paria River entering the Colorado at Lees Ferry contributed only 0.16 percent of the river's volume but added almost 5 percent of the river's sediment load (Iorns and others, 1965; Andrews, 1991). The Paria still carries an average of about 3 million tons of sand, silt, and clay a year into the Colorado. The Little Colorado River, joining the Colorado 62 miles farther downstream, brings in three times more sediment.

In Grand Canyon, the Colorado is a major desert river which once wielded two powerful tools to shape its environment: intermittent high flows and a tremendous supply of sediment. Great volumes of sand were stored along the main channel under conditions of lower flow. During floods, sand would be mobilized from the bottom and draped along higher terraces not normally inundated. As a result, the river deposited beautiful sand beaches throughout the Canyon. Despite shrinkage since the completion of Glen Canyon Dam, these beaches remain an integral aspect of the Canyon — about a third of a million people have used them for camping sites. In the pre-dam landscape, these river deposits were also the substrate for the mesquite, catclaw, and hackberry trees that comprised the flood-level plant community and on which various smaller plants and animals relied.

Confluence of the Little Colorado River at the Colorado River in the Grand Canyon

At 2:00 p.m. on March 13, 1963, the gates of Glen Canyon Dam swung shut. The Colorado River and Grand Canyon would never be the same. Glen Canyon Dam was to be the cornerstone of the Colorado River Storage Project, a series of six dams on the Colorado, Green, San Juan, and Gunnison Rivers (Martin, 1989). Coupled with Hoover Dam, 280 miles downstream, Glen Canyon would help provide flood control, irrigation, and municipal water supply for Arizona, California and New Mexico. Lake Powell, a 26.7-million-acre-foot reservoir created by the dam, would provide recreation for millions of people each year. Sediment retention within Lake Powell would prolong the life of Hoover Dam and its Lake Mead.

The 1956 Federal legislation authorizing Glen Canyon stipulated that electricity would be generated by the dam so as to most efficiently recoup construction costs not only for Glen Canyon but other projects within the Colorado River Storage Project.

The most compelling early impetus for the dam's construction was the fact that each of the states through which the Colorado, San Juan, and Green Rivers flowed were anxious to reserve as much water as possible for their own use. The Colorado River Compact of

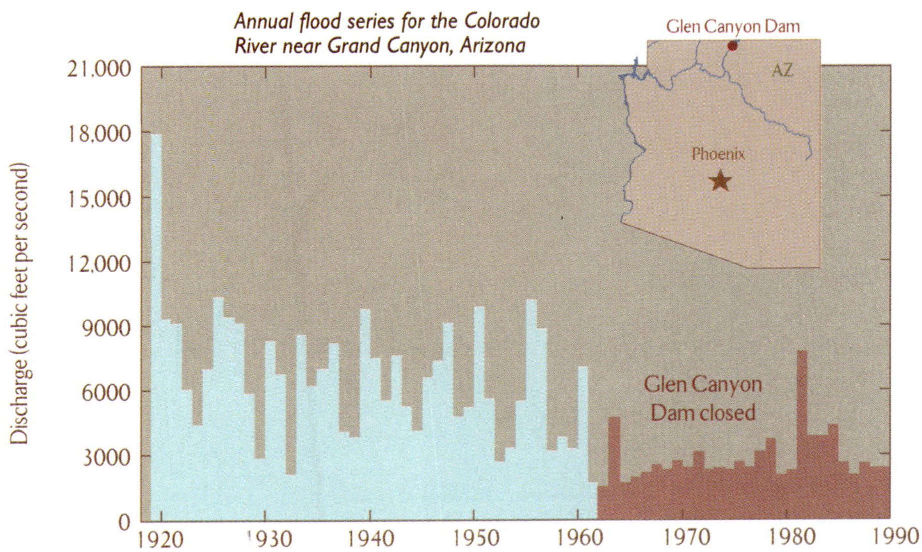

1922 had divided the river among the upper (Wyoming-Colorado-Utah-New Mexico) and lower (Arizona-Nevada-California) basin states. Today, Glen Canyon Dam is operated as the main spigot separating these upper and lower basin states. Its mandate is to release 8.23 million acre-feet a year and not a drop more unless Lake Powell is threatening to spill over (Ingram and others, 1991).

The Colorado River Compact and five subsequent congressional acts are the basis for year-by-year management of water released by Glen Canyon Dam. The U.S. Bureau of Reclamation adjusts the dam's releases on a month-to-month basis according to its projections of how much water is likely to flow into Lake Powell and how much room is left in the lake. The hour-by-hour release of each month's total is, in turn, controlled by the Western Area Power Administration (WAPA), a Federal power broker whose mandate is to sell power to utility companies at the best price WAPA can arrange.

Embedded within Glen Canyon Dam are eight penstocks, each leading to a turbine. The Bureau of Reclamation was originally allowed to send up to 33,200 ft^3/s through these eight penstocks for power generation. Operations are currently restricted by law to a release no greater than 25,000 ft^3/s, resulting in a peak generation capacity of approximately 1,050 mw (U.S. Department of the Interior, 1995). When emergencies arise that require excess water to be released, anything above turbine capacity must be directed through bypass tubes and the spillways, and therefore does not generate electricity. From the perspective of power generation, that water is wasted.

During the eighteen years between dam closure and 1980 when Lake Powell was filling, the Bureau of Reclamation released an average of 8.2 million acre-feet per year

Photographers in the Grand Canyon reproducing photographs taken by Stanton a century before

Repeat Photography

Robert Brewster Stanton was an unlikely visionary. He was an engineer who, in 1889, set out to establish a railroad route that followed the Colorado River, intending to deliver Colorado coal to California markets. In the course of two river trips, he made 445 high-quality, large-format photographs of the river and its banks through Grand Canyon. One hundred years later, Robert Webb of the U.S. Geological Survey followed Stanton's footsteps, precisely relocating the camera position of the original photographs. Webb has analyzed the matched photographs to document changes that occurred during the intervening century (Webb, 1996). In some cases, side canyon floods have carried tremendous quantities of new material down to the river's edge. At other places, even the most precariously balanced small pebbles are resting exactly as they did when Stanton came through. The photographs document the changing nature of the rapids on the Colorado. And they hold significant information about vegetation. Webb has found individuals of 41 plant species still growing exactly where Stanton photographed them in 1890. The photographs document changes in beaches that can be related to the construction of Glen Canyon Dam. Many beaches are deflated, but some have persisted in much the same condition as in 1890. The most severe erosion is closest to Glen Canyon Dam; the effects diminish with distance downstream (Schmidt and others, 1995).

In the winter of 1889-90, Robert Brewster Stanton led the second successful expedition through Grand Canyon. On the morning of February 20, 1890, after camping on a small beach at mile 115, Stanton photographed this view (number 542, National Archives, Washington, D.C.) at 8:00 a.m. while the crew loaded the last of the equipment and supplies onto the 22-foot wooden boats.

Steve Tharnstrom replicated this upstream view on February 22, 1992, at 9:00 a.m. (Webb, 1996). Because of considerable erosion, the sand bar could be used only with considerable difficulty by modern river runners. The sand bars near water level have steadily decreased in size, but the high sand bar on the far right side of the view aggraded as a result of the 1983 flood.

Sockdolager Rapids on the Colorado River in the Grand Canyon

through Glen Canyon Dam — the allocation earmarked for the lower basin states and Mexico. WAPA managed the hour-by-hour release of water by reducing flow through the dam's penstocks at night and increasing flow during the day. It was not unusual to see daily fluctuations during early summer from 8,000 to as much as 28,000 ft³/s. Minimum flows during the weekends were even lower.

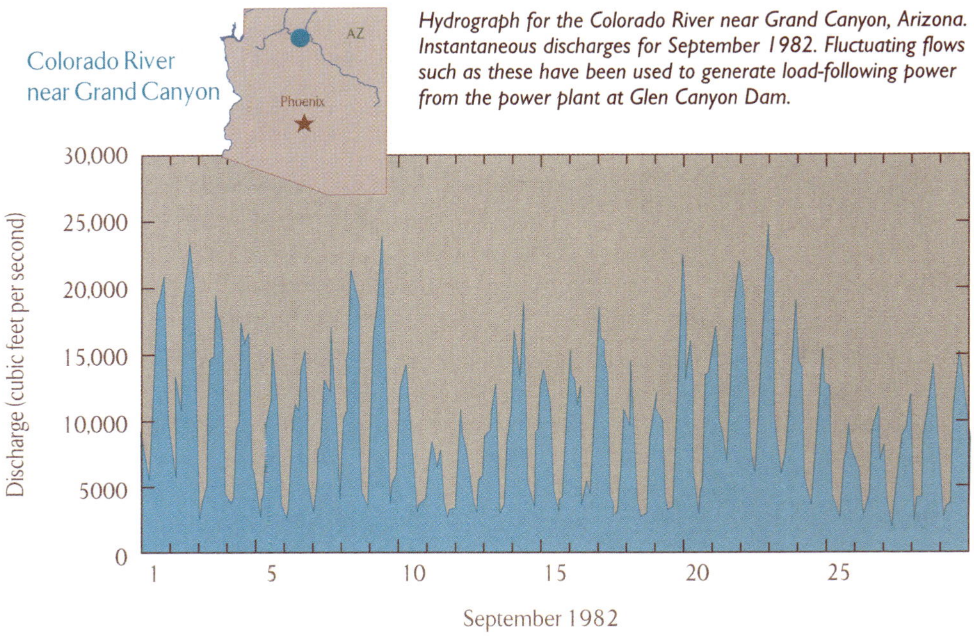

Hydrograph for the Colorado River near Grand Canyon, Arizona. Instantaneous discharges for September 1982. Fluctuating flows such as these have been used to generate load-following power from the power plant at Glen Canyon Dam.

Lake Powell filled for the first time in June 1980, and the Bureau of Reclamation subsequently endeavored to keep the lake as full as possible (Carothers and Brown, 1991). High flows were released during winter and summer peak demands, and the reservoir was intentionally lowered during the winter to accommodate the anticipated inflow of May and June floods. The story of flow regulation at Glen Canyon Dam took a sudden and dramatic turn in 1983. By late June, Lake Powell was again full, and an unexpectedly large runoff roared into Lake Powell. The Bureau and WAPA were releasing maximum penstock flows through Glen Canyon, but the lake still rose at a rate of six inches a day. Plywood splashboards were placed along the top of the spillway gates to hold the water back. Finally, the inevitable happened: the reservoir could no longer hold the inflowing water and the dam's jet tubes and spillways had to be opened. On June 29, 1983, the Colorado River at Lees Ferry was flowing 97,300 ft³/s for the first time since Glen Canyon Dam was completed. This unprecedented discharge was so powerful that cavitation plucked great blocks of concrete and sandstone from the spillway walls (Carothers and Brown, 1991). The dam's structural integrity was threatened but the only failures were inside the spillways.

When it rains, it pours. 1984 was another record-breaking year for runoff in the Colorado River drainage, greater even than 1983. However, due to better planning, the spillways were not needed. The period 1983 to 1986 turned out to be the wettest consecutive four-year period on record. During each of these years, Glen Canyon Dam released an average of 18.5 million acre-feet, far in excess of its target of 8.23 million acre-feet. Instantaneous discharge from Glen Canyon never again reached the high levels of June 1983, but were commonly in the 45,000 to 50,000 ft^3/s range during the late spring and early summers.

One certainly wonders what effect all this hydraulic energy had on Grand Canyon. River runners and scientists in the mid-1970s had noticed that some beaches were disappearing, and that plant and animal life along the river was changing. The Glen Canyon Environmental Studies (GCES) program had been started in 1982 by the Bureau of Reclamation to investigate the dam's effects on downstream environmental and recreational resources (Wegner, 1991). In 1989 the Secretary of the Interior announced that a Glen Canyon Dam Environmental Impact Statement (EIS) would be required for continued operation of the dam. Congress passed the Grand Canyon Protection Act in 1992, stipulating that Glen Canyon Dam is to be operated in a manner that protects resources within Grand Canyon, and that long-term scientific studies be conducted to monitor the downstream effects of the dam.

Guided by these multiple mandates, GCES joined forces with the National Park Service, U.S. Fish and Wildlife Service, Geological Survey, and the Arizona Game and Fish Department to study the Colorado River through Grand Canyon. Together, this alliance of agencies has investigated the regulated river's impact upon fish, birds, insects, and mammals including river runners (U.S. Department of the Interior, 1989). The alliance examined archaeologic sites that lay within the river's potential grasp (Hereford and others, 1991), studied all manner of plants along the riparian corridor (Stevens and Ayers, 1993), and surveyed many types of sediment within Grand Canyon (Webb and others, 1989; Schmidt and Graf, 1990; Beus and Avery, 1992). It is this final impact upon sediment that will receive a closer look here.

Cost of Habitat Maintenance Flows

Many variables contribute to the cost of a beach-building flow. A simplified accounting model can show how to estimate the cost of a habitat-maintenance flow. Assuming a full Lake Powell, 40.78 kW are generated for each cubic foot per second released by the dam under 33,200 ft^3/s; any water in excess of this discharge cannot be used to generate power. If 50,000 ft^3/s was released for one week, 20.08 billion ft3 would have passed through the dam's turbines and 10.16 billion ft^3 would have bypassed the turbines.

In 1994 dollars, power produced during peak demand (7 a.m.-11 p.m., Monday through Saturday) has a value of $0.05 per kilowatt-hour (kw-hr), and $0.0137/kw-hr during off-peak periods (Ron Moulton, Western Area Power Administration (WAPA), oral commun., 1993). During a beach-building flow, WAPA would produce $7,834,000 worth of power. We now assume that the alternative is "typical" fluctuating flow; for our model, flow will either be 20,000 or 8,000 ft^3/s for peak and off-peak hours respectively. WAPA would have produced $4,237,000 worth of power during this hypothetical week — and still have had 13.3 billion ft^3 left in Lake Powell to be released later. If it was possible to release the entire 13.3 billion ft^3 at peak-demand times, that water would produce power worth $7,503,000.

During a beach-building flood, WAPA would have received $7,834,000. Alternatively, WAPA would have received $4,236,683 if the dam had been operated during that same week according to a typical fluctuating-flow regime. But water worth as much as $7,503,000 would still be stockpiled in Lake Powell. All together, the beach-building flow would have cost WAPA $3,905,196. But Glen Canyon Dam produces $140,000,000 in revenues per year. If a habitat-maintenance flow occurred once every five years, it would represent about half of one percent of the gross receipts from operations of the dam.

WAPA, GCES, and other interested parties have created more sophisticated models that integrate variables such as flow duration, water availability, pricing structure, and alternative generation costs for a beach-building flow. These more sophisticated estimates place the cost of a beach-building flood between $500,000 and $1,500,000 (David Wegner, GCES, oral commun., 1993). Such a price tag must be weighed against the flood's possible benefits to both the multi-million dollar rafting industry and the environmental quality of Grand Canyon National Park as mandated by the Grand Canyon Protection Act.

Palisades Creek in Grand Canyon

Water released by Glen Canyon Dam carries almost no sediment Any sediment along the Colorado River below the dam was either already present when the dam was built, or has since been brought in by tributaries. Tributaries deliver sediment to the river in one of two basic patterns slow steady additions of silt and sand, or sporadic floods containing everything from huge boulders to fine silt. Both processes are important to the evolution, homeostasis, and general well - being of Grand Canyon.

Time series of changes in the bed elevation of the Colorado River at Lees Ferry, Arizona, 1921-1986 (modified from Burkham, 1986). Before construction of Glen Canyon Dam, the bed was annually scoured 20-30 feet during spring floods. After June 1965, in the absence of river-borne sediment, bed elevation fell and did not recover its previous elevation. Now releases from Glen Canyon Dam have little effect on the armored bed.

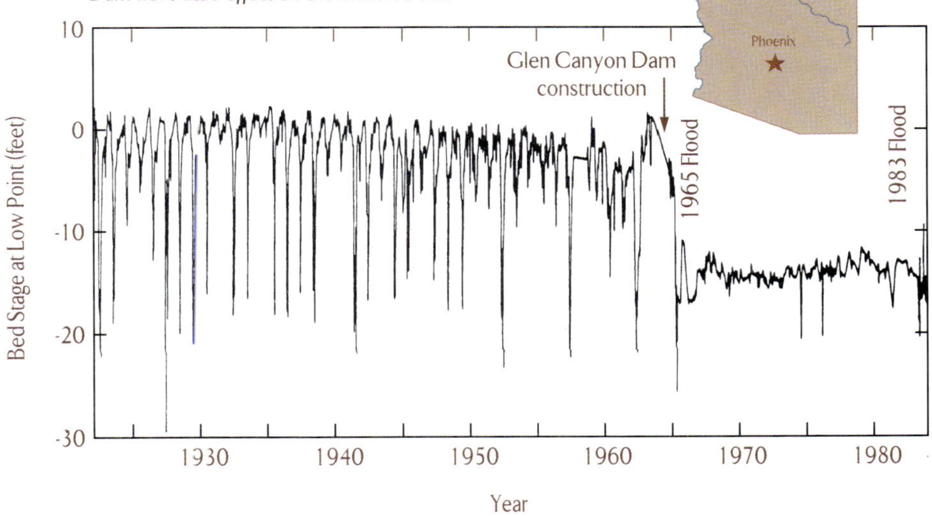

Within Grand Canyon, the Paria and Little Colorado River provide the lion's share of the total volume of sand brought to the mainstream river in the post - dam era. Debris flows are another important source of materials deposited in the Colorado River. A debris flow consists of cobbles and boulders within a matrix of sand and silt that has been transformed into a fluid with as little as 10 percent (but no more than 40 percent) water by volume. Hydrologists with the U.S. Geological Survey have examined debris flows that sweep down steep tributary canyons within Grand Canyon (Webb and others, 1989). These slurries usually start as rain - induced slope failures of clay - rich shales or talus, part way up the walls of Grand Canyon. The falling detritus hits the bottom of a tributary canyon, entraining cobbles and boulders enroute, and then surges on down to the river with velocities estimated at 10 to 30 ft/s (Melis and others, 1994).

An unusually heavy rainstorm inundated the upper drainage of Crystal Creek in December 1966. The resulting debris flow had a discharge of about 10,000 ft^3/s of water, mud, and boulders. The slurry surged well out into the Colorado River, severely constricting the river's channel against the opposite wall. Boulders deposited in the main channel instantly transformed Crystal from a benign riffle to one of the Canyon's most difficult rapids. Indeed, virtually all the rapids in Grand Canyon have been created by debris flows hurling a fan of boulders into the river. Such events recur at intervals of as little as one every 20 years to less than one per century for individual tributary canyons (Melis and others, 1994).

Debris flows beget rapids because the boulders they transport are too large to be quickly removed by typical discharges of the Colorado River (Graf, 1980). The high river flows of 1983 did partially rework the Crystal Creek debris fan, moving boulders and changing some aspects of the rapid. U.S.G.S. geologist Susan Kieffer estimated that a flood of 400,000 ft^3/s would be required to remove that debris fan (Kieffer, 1985). Certainly a 300,000 ft^3/s flood, like the one that drove Jerry Johnson's cat up a tree in 1884, would significantly alter or even smooth out many of Grand Canyon's rapids. But smaller floods may also help maintain the river's channel; the deposits of a larger debris flow at Lava Falls Rapid were removed by a 120,000 ft^3/s flood in 1941 and dam releases augmented with small floods from the Little Colorado River have eroded several debris fans (Melis and others, 1994). A question relevant to today's flood-control policy at Glen Canyon Dam is how many debris flows will occur before the next large boulder - rolling flood sweeps down the mainstem Colorado River.

Island at the mouth of the Little Colorado River

Debris fans constrict the Colorado River, setting up the conditions that create sand bars. Water flow accelerates within the constriction created by the debris fan. Downstream from the fan, the river expands and decelerates. This variation of both channel shape and velocity spawns an eddy, with its water flowing upstream, opposite the direction of the main channel. Water within an eddy tends to have a velocity well below that of the rest of the river. As water (with whatever sand it happens to be carrying) moves from the main channel into the eddy, it decelerates and consequently drops some of its suspended sediment. Sand begins to fill the eddy.

Schmidt and Graf (1990) built upon the work of Howard and Dolan (1981) to develop a descriptive model of sand that is deposited in eddies. They noticed that some sand is deposited just below the point where the river's main current separates from the channel walls, forming a separation bar. More sand is deposited downstream where the current reattaches to the channel walls, forming a reattachment bar. As sand is deposited or removed, and as the river level rises or falls, the two bars may shift slightly up or downstream, but they do not migrate out of the eddy in which they form. Schmidt and Graf also mentioned channel margin bars, which generally do not occur in association with eddies, but along the river banks where flow is slower than in mid - channel.

Sand within the Colorado River is cached in one of four locations that can be compared to bank accounts on the main channel bed, underwater within an eddy, perched above the present water line in the form of beaches, or in transport by the moving river water. We can measure beaches and measure suspended load, but at this point we can only guess the relative sizes of underwater eddy deposits and channel - bed deposits. Experience suggests that eddies can trap tremendous quantities of sand. People camping along the Colorado River are most likely to use beaches that formed as separation bars when the river flowed at greater than 30,000 ft^3/s.

The river can rise and overtop a debris fan, and if this occurs, the flow is likely to erode rather than deposit a beach. Now sand is being shuffled from one account to the other at a furious pace. The Colorado River or any canyon river is always in a state of dynamic equilibrium. At a given discharge, it will be creating with one hand and destroying with the other. Change the discharge, change the suspended sediment, change the channel shape and the equilibrium will shift all along the river.

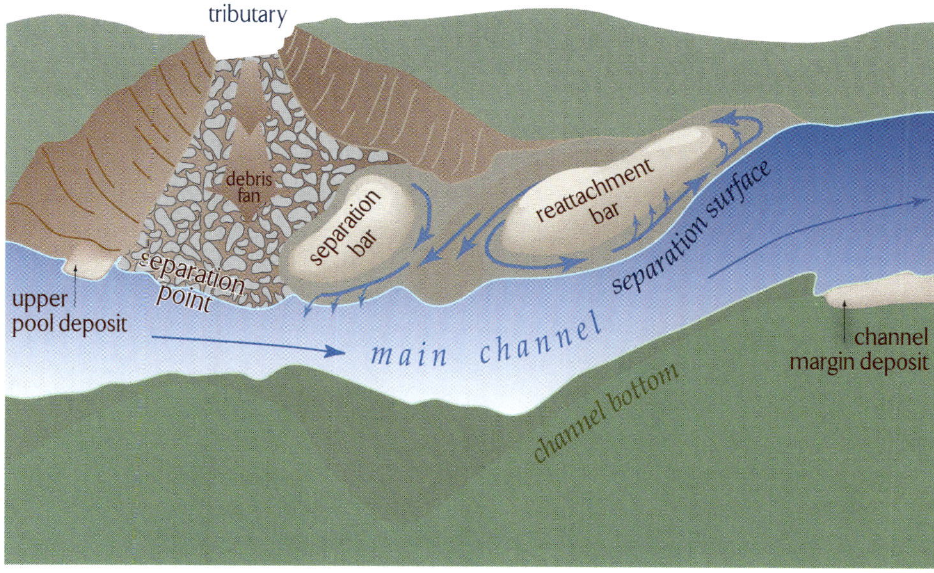

Sand bars are draped around an eddy in a predictable pattern in most bedrock-controlled rivers. The debris fan, which is formed from periodic debris flows from the tributary canyon, constricts the river. The expansion zone downstream allows flow separation in an eddy. Separation bars form on the downstream side of the debris fan, beginning at the point of flow separation. Reattachment bars form at the stagnation point on the downstream side of the eddy. Channel margin bars are deposits that form along the channel margin in the lee of boulders and other obstructions (from Schmidt and Graf, 1990).

Many Grand Canyon beaches are smaller now than they were before Glen Canyon Dam was built (Kearsley and others, 1994). River - runner suspicions and GCES studies have long focused on fluctuating flows as the prime culprit responsible for the disappearance of beach sand. GCES scientists have taken great pains to examine the rate at which daily fluctuations occur, studying the effects of the range of daily fluctuations as more and then less water is released from the dam for peak - power generation. The scientists have

The Colorado River in Grand Canyon

also studied the possibility that when Glen Canyon Dam rapidly decreases its release of water (a fast "downramping" rate), beaches downstream would erode more quickly (Beus and Avery, 1992).

Fluctuations in flow do cause erosion of beaches, and beaches will continue to disappear in Grand Canyon in the absence of sediment - laden flood flows. Gravity will eventually pull beach sand back underwater, and the river will carry the sand downstream. At best, some dam operation regimes might be more successful than others at minimizing beach erosion. But the moderate daily high flows associated with the generation of hydroelectric power at Glen Canyon Dam are not likely to rebuild the high beaches that once existed in the canyon.

Should river runners despair? Perhaps not. Although scientists now understand that beach erosion is an inevitable process, and despite the fact that most sediment once carried by the Colorado River is now stored behind Glen Canyon Dam, there may still be sufficient sediment supplied by the tributaries to replenish Grand Canyon's beaches at the rate of once every few years. The secret to this replenishment lies primarily with the sand supplied by the Paria and Little Colorado rivers.

Imperial Dam north of Yuma Arizona

Beach at Badger Rapid in Grand Canyon

As early as 1981, University of Virginia researchers Alan Howard and Robert Dolan estimated that the bed of the Colorado River was accumulating sediment despite the observed erosion of camping beaches, because the rate of sediment contribution from the Paria and Little Colorado exceed the transport capacity of the flood-regulated Colorado. Under the auspices of the GCES, Wilson (1986) made a river-length side scan sonar image of the river bottom and found that substantial parts of the channel bed were covered with sand. Randle and others (1993) estimated that more than 25 million tons of sand accumulated along the river bed in the first ninety miles of Grand Canyon between 1965 and 1982.

Beaches disappearing on the channel margins but sand accumulating on the channel bed. This is not a contradiction. Two processes were operating at the same time: beaches slumping into the river, and tributaries like the Paria simultaneously conveying new sediment to the river. Until 1983, the sediment-carrying capacity of water released from Glen Canyon Dam was insufficient to mobilize sediment accumulating on the riverbed throughout Grand Canyon. Other than one brief burst of high water in 1965, the canyon had not experienced a truly big flood since dam closure. But during the highwater years between 1983 and 1986, the floods came and the river's sediment account was shuffled in some places, squandered in others. Some sand was transferred to higher elevations in eddies, but most sand was stripped from the river bed and carried downstream.

This loss of sediment occurred in two stages. When the first high releases of June 1983 rolled through the canyon, the 97,300 ft^3/s discharge stirred up sand from the main channel bed, from eddies and from beaches that had normally been above high water line. This frothy brown mixture rushed downstream and in places was redeposited as beach sediment that remained after that first flood abated. Brian and Thomas (1984) concluded that, on average, beaches in the first 180 miles of Grand Canyon suffered net erosion in 1983, but beaches farther down gained sediment. Overall during this first stage, the canyon lost a lot of sediment. After the flood receded, however, a small but significant percentage of available sand remained perched where it had been deposited by the high waters of the flood. This sand formed new beaches or added to preexisting ones.

The second stage of sediment loss began in 1984 when 45,000 ft^3/s flows were released three summers in a row. With each release, the riverbed had significantly less sediment to mix into the water. With each release, more sediment was stripped from the canyon and less was returned to beaches after the flooding ceased. By 1986, the beaches of Grand Canyon were scoured, and the riverbed had little material with which to rebuild them.

Grand Canyon National Park, Arizona

It's all too easy to second-guess the engineers at the Bureau of Reclamation, saying that they should have operated Glen Canyon Dam some other way. But the period between 1983 and 1986 were anomalously wet years. The Bureau of Reclamation has since unofficially designated the uppermost half-million acre-feet of storage in Lake Powell as a buffer against floods, in an attempt to avoid the reservoir spills of the mid-1980s.

Smillie and others (1992) developed a sand budget that suggests that if daily fluctuating flow is kept between 5,000 and 20,000 ft^3/s, only 484,000 tons of sand will be transported out of the canyon in a given year. Contrast that figure with the 1.2 million-ton annual sand input of the Paria River alone. Thus, sand can be stored on the channel bed and, when a sufficient amount has accumulated, could be remobilized by intentional high floods to rebuild beaches farther downstream. The EIS for the operation of Glen Canyon Dam (U.S. Department of the Interior, 1995) called for flows to be held between 5,000 and 25,000 ft^3/s. Presumably, sediment will accumulate on the riverbed under those conditions.

With increasing frequency, scientists have called for "beach-building flows" (Schmidt, 1992) or "habitat-maintenance flows" (U.S. Department of Interior, 1994). Beginning on March 26, 1996, Glen Canyon Dam spilled 45,000 ft^3/s for eight days – the first intentional flood ever released for environmental purposes. The Grand Canyon crawled with scores of scientists, measuring the flood's effects on beaches and backwaters, photographing eddy patterns and sampling sediment loads. When the flood receded, a great deal of clean new sand had been perched well above the normal high-water line. The long beautiful beaches were reminiscent of the pre-dam Grand Canyon.

The Bureau of Reclamation picked up the tab for the 1996 flood, estimated to be as much as $4.25 million dollars (Dave Sabo, personal commun., 1996). This covered the cost of lost revenues and scientists salaries. Was it worth it? Time will tell. The beaches quickly began to readjust when the water dropped back to normal flows. Studies of the flood's biologic impacts will take months to complete. But a tremendous precedent had been set: never before had a flood been used as a management tool.

The Grand Canyon experienced larger floods during the 1980s but those occurred in quick succession and in response to unusually large inflow into Lake Powell. Beach-building flows must be released only after adequate amounts of sand have built up on the river bed. Then high flows are most likely to be beneficial. Intentional floods might be timed to coincide with natural floods down the Paria and Little Colorado rivers, which would carry a wider array of sediment sizes into eddies. Or the Bureau of Reclamation might be given the flexibility to time such releases to coincide with predicted high levels in Lake Powell, reducing the threat of an uncontrolled spill.

Periodic beach-building flows are an exciting new tool in dam management because they are an outgrowth of scientific investigations into the downstream effects of dams. At this point, no hydrologist or geologist pretends to know the ideal parameters for such a flow: how big, how long, how often. What we do know is that a beach-building flow may benefit some resources within the canyon while simultaneously degrading others. Some beaches would be enlarged, others would shrink. What would the effects be on native fish, on archaeologic sites, or on riparian vegetation? How much revenue would be lost because of bypassed electrical generation? Could the benefits of such a flow be used to balance the environmental cost of fluctuating flows? Ultimately dam management must aspire to a balance between environmental benefits and societal costs. In Grand Canyon, much hangs in the balance: the environmental quality of one of the nation's premier national parks on one side, and the power-generation capability of one of the West's largest hydroelectric power plants on the other.

Rapid in Grand Canyon National Park, Arizona

The Role of Science

Science can play a key role in the management of dams to minimize downstream impacts. Research provides hard data on environmental changes that might occur if dam releases ar altered. Modeling can be used to optimize the management plan for a dam-river system. But scientists do not — an should not — make the final decision about the "best" management of a dam. Society must decide what it expects of its dams, and what it needs of its rivers.

Few large dams will ever again be built in the United States.

It is certainly reasonable to expect that the downstream impacts of any future dam would be closely scrutinized before construction. The real task facing scientists (and society) is to examine the effects of dams already in place. Although some dams such as Glen Canyon have already received considerable attention for their downstream impacts, the effects of most dams have not been studied at all.

This Circular has surveyed a few rivers to illustrate some of the adverse effects that dams can create downstream. The undammed upper Salt River determines its own capacity for transporting sediment, and continues to shape a unique environment that supports natural vegetation and native birds and fish. In contrast to the Salt, dams on the

Elephant Butte Dam near Truth or Consequences, New Mexico

Sandbar on the Salt River

The Rio Grande ... Mesilla Dam nea... Cruces, New Me...

Snake River are operated for peak power production, and the river downstream has lost a significant amount of its sediment and may lose its native salmon population. The Rio Grande is little more than an inconvenient afterthought below Elephant Butte Dam; the limited flows below the dam are incapable of transporting sediment that now builds up in the channel. Buford Dam has significantly changed the character of the Chattahoochee River; the river's flow represents a compromise between the needs for hydropower and water delivery, and Atlanta's needs for recreation. The Platte River has steadily shrunk in flow and channel width over the past century and a half; persistence of the sandhill crane may require that the original variability and viability of the system be restored. Natural environments of the Green and Colorado Rivers developed in a closed system defined by seasonal variability; today dams control these rivers, and several native fish have become endangered or extinct as a result.

Flaming Gorge Dam on the Green River in Utah

The application of science to the study of downstream effects is new and growing. The questions that society is now asking can be complex in some cases, simple in others. For some rivers where impacts have not been studied, it may be possible to efficiently address questions about downstream problems by applying information and insights gathered on similar rivers. In other cases, it may be necessary to undertake new research. Some questions may be sufficiently complex that answers will require sophisticated and costly new research and development of predictive tools. In other cases, we will be able to rely on already-existing data collected before and after construction of a dam.

Dams have profound but varied impacts on the rivers which they harness. A pattern of seasonal high water is usually replaced by flow that resembles a daily tidal flood. Sediment typically is trapped in the reservoir above a dam. Downstream, sediment may either diminish or accumulate, depending on distance from the dam, input from tributaries, and the difference between pre-dam flows and dam releases. The channel downstream from a dam may narrow in some reaches or widen in others. Riparian plants may increase or decrease in response to flood control and the altered amount of flow. Water temperature, suspended sediment, and flow patterns may change sufficiently that fish – particularly native fish – face an entirely different set of challenges to which they must adapt.

All other things being equal, dams of different design and operation will produce dissimilar effects downstream. The hydrograph created by a

Dredging operation on the Chattahoochee River in Duluth, Georgia

Dike reinforcement on the Rio Grande at Redford, Texas

dam that produces load-following power will look totally different than one from a dam whose reservoir provides base-load power. Flood-control operations have different impacts than run-of-the-river releases. Dams operated for water-supply or irrigation purposes have different effects than hydropower dams.

We cannot responsibly expect dams of one design to be operated according to the demands of quite another design. For example, the small reservoirs and high flows associated with the Hells Canyon Complex dams do not allow these dams to be efficiently operated as primary storage facilities or for flood control. Retrofitting is always a possibility, but the cost of structural changes can be astronomical.

Scientists are increasingly being called upon to suggest dam operation regimes that will minimize a particular negative impact to the downstream environment. Geoscientists have a large role to play because many of the adverse impacts relate to changes in sediment transport and physical aspects of the river's channel. Scientific insights gained from geomorphic research may be useful, but these suggestions must be integrated with recommen-

Broad Canyon Dam, a sediment retention structure, upstream from the Rio Grande near Las Cruces, New Mexico

Cottonwoods in Pueblo Colorado Wash near Ganado, Arizona

The Colorado River below Westwater Canyon, Utah

dations from the fields of biology, economics, and engineering. Any scientific recommendation that alters a dam's operation must be evaluated in the context of the people whose lives will be most directly affected – people whose interests may be as disparate as the river-runner who wants steady in-stream flow, and the farmer trying to irrigate his crops with low-salinity water.

The North Platte River below Keystone Dam in North Platte, Nebraska

Floods are a key element in the future management of dams. Without periodic high flows, some channels downstream from dams will aggrade with sediment or narrow with overgrown vegetation. Two or three flood-free decades may have been traded for more devastating floods in the future. As exemplified by the Hells Canyon Complex, not all dams are designed to prevent floods. The unpredicted 1983 high releases in Grand Canyon show that a fickle climate can defeat the best efforts at flood control. Dredging, channelization, and mowing riparian vegetation are temporary measures that can forestall problems caused by reduced floods; but in some cases, like the Rio Grande below Elephant Butte Dam, our interventions have not always been timely or effective.

Smokey the Bear has fallen out of favor in the last few years. Decades of fire suppression have radically changed the makeup of American forests, and now scientists and the public are questioning the wisdom of long-term fire suppression. Analogously, we are just beginning to realize that flood suppression has created situations along some rivers that may be responsible for future problems or disasters. For most rivers, floods are an integral element of their natural equilibrium, necessary for maintaining channels, and replenishing bankside sediment and nutrients.

The benefits of dams are tightly woven into our daily lives. Despite the problems that they can create, most dams will not, and should not, simply be dismantled. Part of learning to live with them is to realize the extent of adverse effects, and to appreciate that those effects can be manipulated to some degree. Any number of tinkerings in dam operations can be proposed to minimize downstream changes – low fluctuating releases, seasonally-adjusted flows, periodic high floods, sediment augmentation, multilevel intake structures. But the fact is, with a dam in place, we forever forego an entirely natural environment below the dam. As we try to maximize one quality within a downstream riparian system, we will inevitably alter and possibly degrade other qualities. There can be only one river and one flow regime below a dam. To pretend otherwise, to promise more than can be delivered, is naive at best, irresponsible at worst.

Any regulated river can respond to its dam with a dizzying array of changes. Future management of dams and rivers must take this dynamic nature into account. Adaptive management attempts to track these changes through time. As dam operations are modified, the resultant downstream effects are tracked and integrated back into the dam's management plan. With this approach, the very uncertainty of river response is used as an advantage in the development of monitoring programs and dam management alternatives.

Scientists have already spent a great deal of time and money studying rivers below dams, $90 million alone at Grand Canyon since 1982. Have we accomplished anything? Was the money well spent? Not surprisingly, most work on the downstream effects of dams has been retrospective: scientists examining changes after they happen. The knowledge gained from this sort of science is valuable, but payoffs are typically discussed in the future tense. Times are changing. Flaming Gorge Dam is being run according to the Fish and Wildlife Service's Biological Opinion, and Glen Canyon Dam operations have been altered by the scientific findings of the Glen Canyon Environmental Studies program. The worlds first intentional flood roared through Grand Canyon in 1996. Scientists must soon assume the uncomfortable responsibility of looking back over their shoulders to see if the changes they induced actually saved an endangered species, reconstructed a beach, or improved a habitat.

We should not confuse the roles of scientists with those of engineers or politicians. Nevertheless, no one works in a vacuum. Our ideas and observations will be put to test, not just in the rarified atmosphere of academic science, but integrated into the very real world that runs according to cost per kilowatt - hour, water rights, and mandated protection of the natural environment. Scientists dealing with the downstream effects of dams — at levels ranging from basic research to applied engineering — must formulate questions whose answers can ultimately make a difference. To strive for anything less is to be just another bureaucrat.

Luna Leopold, one of the nation's leaders of water research and resource issues, concluded a symposium on Colorado River ecology and dam management by stating:

> *The interpretation of scientific understanding, however limited or conditional, must be translated by the scientists into concrete recommendations designed for the manager. We must make small changes initially. But it should be understood very clearly by the people who are supporting the investigation that more changes will be needed. The need for long-term measurement is the unassailable conclusion of the studies made to date. The data-collection program should be designed with great care, it should consider a wide variety of data needs, and each part should be in stalled and operated as soon as practicable even though not all parts...begin at one time* (Leopold, 1990).

Leopold's comments go beyond issues restricted to the Colorado River in Grand Canyon. As we have seen, scientists have already addressed various aspects of the downstream effects of dams on many rivers. The US Geological Survey has a long history of monitoring and research concerning the nation's streams. Scientists within the Geological Survey and other Federal agencies are building upon long-term datasets of gaging trends to develop sophisticated explanations of the movement of water and sediment below dams.

Rivers and riparian habitats are critical elements of our environment. As a society, we have gradually refined our vision of how we would like our rivers to look. We have tried to achieve a balance between extractable resources and environmental values. We need a concerted effort to understand the changes that dams can cause and learn to predict the changes that are most likely to occur. Some useful management tools have been proposed: intermittent high releases from dams may have beneficial downstream effects. Science can offer options for better dam management; it's up to society to choose.

Aldridge, B.N., and Hales, T.A., 1984, Floods of November 1978 to March 1979 in Arizona and west-central New Mexico: U.S. Geological Survey Water-Supply Paper 2241, 149 p.

Andrews, E.D., 1986, Downstream effects of Flaming Gorge Reservoir on the Green River, Colorado and Utah: Geological Society of America Bulletin, v. 97, p. 1012-1023.

Andrews, E.D., 1991, Sediment transport in the Colorado River basin, *in* Colorado River Ecology and Dam Management: Washington, D.C., National Academy Press, p. 54-74.

Andrews, E.D., and Nelson, J.M., 1989, Topographic response of a bar in the Green River, Utah, to variation in discharge, *in* Ikeda, S., and Parker, G. (editors), River Meandering: American Geophysical Union, Water Resource Monograph 12, p. 463-485.

Atlanta Regional Commission, 1991, Atlanta Regional Water-Supply Plan: Atlanta, Georgia, Atlanta Regional Commission, 62 p.

Atlanta Regional Commission, 1992, Atlanta Region Outlook, Executive Summary: Atlanta, Georgia, Atlanta Regional Commission, 34 p.

Baker, V.R., 1977, Stream-channel response to floods, with examples from central Texas: Geological Society of America Bulletin, v. 88, p. 1057-1071.

Bartlett, R.A., 1984, Rolling Rivers, An Encyclopedia of America's Rivers: New York, McGraw-Hill Book Company, 398 p.

Beus, S.S., and Avery, C.C., 1992, Influence of variable discharge regimes on Colorado River sandbars below Glen Canyon Dam: Flagstaff, Arizona, U.S. Department of the Interior, National Park Service Cooperative Agreement #CA 8006-8-0002, 59 p.

Brian, N.J., and Thomas, J.R., 1984, 1983 Colorado River beach campsite inventory, Grand Canyon National Park, Arizona: Grand Canyon, Arizona, Grand Canyon National Park, Division of Resources Management and Planning, unpublished report, 56 p.

Burkham, D.E., 1972a, Hydraulic effects of changes in bottom-land vegetation on three major floods, Gila River in southeastern Arizona: U.S. Geological Survey Professional Paper 655-J, 14 p.

Burkham, D.E., 1972b, Channel changes of the Gila River in Safford Valley, Arizona, 1846-1970: U.S. Geological Survey Professional Paper 655-G, 24 p.

Burkham, D.E., 1986, Trends in selected hydraulic variables for the Colorado River at Lees Ferry and near Grand Canyon for the period 1922-1984: Springfield, Virginia, U.S. Department of Commerce, National Technical Information Service, Report PB88-216098, GCES/07/87, 58 p.

Carothers, S.W., and Brown, B.T., 1991, The Colorado River through Grand Canyon—Natural History and Human Change: Tucson, University of Arizona Press, 235 p.

Carrey, J., Conley, C., and Barton, A., 1979, Snake River of Hells Canyon: Cambridge, Idaho, Backeddy Books, 399 p.

Cherry, R.N., Faye, R.E., Stamer, J.K., and Kleckner, R.L., 1980, Summary of river-quality assessment of the upper Chattahoochee River basin, Georgia: U.S. Geological Survey Circular 811, 47 p.

Cullen, A.H., 1962, Rivers in Harness —The Story of Dams: Philadelphia, Pennsylvania, Chilton Books, 175 p.

Dolan, R., Howard, A D., and Gallenson, A., 1974, Man's impact on the Colorado River in Grand Canyon: American Scientist, v. 62, p. 392-401.

Eley, T.J., and Watkins, T.H., 1991, In a sea of trouble—The uncertain fate of the Pacific salmon: Wilderness, Fall 1991, p. 20-26

Eschner, T.R., Hadley, R.F., and Crowley, K.D., 1983, Hydrologic and morphologic changes in channels of the Platte River basin in Colorado, Wyoming, and Nebraska—A historical perspective: U.S. Geological Survey Professional Paper 1277-A, 26 p.

Everitt, B.L., 1993, Channel responses to declining flow on the Rio Grande between Ft. Quitman and Presidio, Texas: Geomorphology, v. 6, p. 225-242.

Faye, R.E., Carey, W.P., Stamer, J.K., and Kleckner, R.L., 1980, Erosion, sediment discharge, and channel morphology in the upper Chattahoochee River basin, Georgia: U.S. Geological Survey Professional Paper 1107, 85 p.

Grabowski, S.J., and Hiebert, S.D., 1989, Some aspects of trophic interactions in selected back waters and the main channel of the Green River, Utah: Denver, Colorado, Bureau of Reclamation, Final Report.

Graf, W.L., 1978, Fluvial adjustments to the spread of tamarisk in the Colorado Plateau region: Geological Society of America Bulletin, v. 89, p. 1491-1501.

Graf, W.L., 1980, The effect of dam closure on downstream rapids: Water Resources Research, v. 16, no. 1, p. 129-136.

Graf, W.L., 1993, Landscapes, commodities and ecosystems—The relationship between policy and science for American rivers, *in* Water Science and Technology Board, Sustaining Our Water Resources: Washington, D.C., National Academy Press, p. 11-42.

Grams, P.E., 1991, Degradation of alluvial sand bars along the Snake River below Hells Canyon Dam, Hells Canyon NRA, Idaho: Middlebury, Vermont, Middlebury College, unpublished Senior Paper, 98 p.

Hereford, Richard, Fairley, H.C., Thompson, K.S., and Balsam, J.R., 1991, The effect of regulated flows on erosion of archaeologic sites at four areas in eastern Grand Canyon National Park, Arizona, a preliminary analysis: U.S. Geological Survey Administrative Report, 58 p.

Hirsch, R.M., Walker, J.F., Day, J.C., and Kallio, R., 1990, The influence of man on hydrologic systems, *in* The Geology of North America, Vol O-1, Surface Water Hydrology: Boulder, Colorado, Geological Society of America Special Publication, p. 329-359.

Holden, P.B., 1979, Ecology of riverine fishes in regulated stream systems with emphasis on the Colorado River, *in* Ward, J.V., and Stanford, J.A. (editors), The Ecology of Regulated Streams: New York, Plenum Press, p. 57-73.

Howard, Alan, and Dolan, Robert, 1981, Geomorphology of the Colorado River in the Grand Canyon: Journal of Geology, v. 89, p. 269-297.

Howe, W.H., and Knopf, F.L., 1991, On the imminent decline of Rio Grande cottonwoods in central New Mexico: Southwestern Naturalist, v. 36, no. 2, p. 218-224.

Ingram, Helen, Tarlock. A.D., and Oggins, C.R., 1991, The law and politics of the operation of Glen Canyon Dam, *in* Colorado River Ecology and Dam Management: Washington, D.C., National Academy Press, p. 10-27.

International Boundary and Water Commission, 1989, Flow of the Rio Grande and related data: U.S. Department of State, International Boundary and Water Commission Water Bulletin Number 59, 137 p.

Iorns, W.V., Hembree, C.H., and Oakland, G.L., 1965, Water resources of the upper Colorado River basin: U.S. Geological Survey Professional Paper 441, 370 p.

Jarrett, R.D., and Malde, H.E., 1987, Paleodischarge of the late Pleistocene Bonneville flood, Snake River, Idaho, computed from new evidence: Geological Society of America Bulletin, v. 99, p. 127-134.

Jobson, H.E., and Keefer, T.N., 1979, Modeling highly transient flow, mass, and heat transport in the Chattahoochee River near Atlanta, Georgia: U.S. Geological Survey Professional Paper 1136, 41 p.

Jones, M.L., and Seitz, H.R., 1980, Sediment transport in the Snake and Clearwater rivers in the vicinity of Lewiston, Idaho: U.S. Geological Survey Water-Resource Investigations Report 80-690, 179 p.

Kearsley, L.H., Schmidt, J.C., and Warren, K.D., 1994, Effects of Glen Canyon Dam on Colorado River sand deposits used as campsites in Grand Canyon National Park, USA: Regulated Rivers Research and Management, v. 9, p.137-149.

Kieffer, S.W., 1985, The 1983 hydraulic jump in Crystal Rapid—Implications for river-running and geomorphic evolution in the Grand Canyon: Journal of Geology, v. 93, p. 385-406.

Krapu, G.L., and Eldridge, Jan, 1984, Crane River: Natural History, v. 93, no. 1, p. 68-74.

Krapu, G.L., Facey, D.E., Fritzell, E.K., and Johnson, D.H., 1984, Habitat use by migrant sandhill cranes in Nebraska: Journal of Wildlife Management, v. 48, no. 2, p. 407-417.

Krapu, G.L., Reinecke, K.J., and Frith, C.R., 1982, Sandhill cranes and the Platte River: Transactions of the 47th North American Wildlife and Natural Resources Conference, p. 542-552.

Lagasse, P.F., 1980, An assessment of the response of the Rio Grande to dam construction—Cochiti to Isleta reach, Albuquerque, New Mexico: Albuquerque, New Mexico, U.S. Army Corps of Engineers, Technical Report, 133 p.

Lagasse, P.F., 1981, Geomorphic response of the Rio Grande to dam construction: New Mexico Geological Society, Special Publication Number 10, p. 27-46.

Leopold, L.B., 1990, Closing remarks, *in* Colorado River Ecology and Dam Management: Washington, D.C., National Academy Press, p. 254-257.

Leuchtenburg, W.E., 1953, Flood control politics—The Connecticut River Valley problem, 1927-1950: Cambridge, Massachusetts, Harvard University Press, 339 p.

Martin, Russell, 1989, A Story that Stands like a Dam: New York, Henry Holt and Company, 354 p.

Melis, T.S., Webb, R.H., Griffiths, P.G., and Wise, T.J., 1994, Magnitude and frequency data for historic debris flows in Grand Canyon National Park and vicinity, Arizona: U.S. Geological Survey Water-Resources Investigations Report 94-4214, 285 p.

Miller, R.R., 1959, Origin and affinities of the freshwater fish fauna of western North America: Zoogeography, v. 41, p. 187-222.

Minckley, W.L., and Brown, D.E., 1982, Wetlands: Desert Plants, v. 4, p. 222-287.

Minckley, W.L., and Clark, T.O., 1984, Formation and destruction of Gila River mesquite bosque community: Desert Plants, v. 6, p. 23-30.

Minckley, W.L., and Deacon, J.E., 1991, Battle Against Extinction: Tucson, University of Arizona Press, 517 p.

Montgomery, Robert, 1991, Our endangered waters: Living Waters, Endangered Waters (Anglers for Clean Water, Inc.), v. 2, no. 1, p. 4-26.

Mueller, J.E., 1975, Restless River: El Paso, Texas Western Press, 153 p.

O'Connor, J.E., 1993, Hydrology, hydraulics, and geomorphology of the Bonneville flood: Geological Society of America Special Paper 274, 83 p.

O'Connor, J.E., Ely, L.L., Wohl, E.E., Stevens, L.E., Melis, T.S., Kale, S., and Baker, V.R., 1994, A 4500-year record of large Colorado River floods in the Grand Canyon, Arizona: Journal of Geology, v. 102, p. 1-9.

Ortiz, A., 1979, Handbook of North American Indians, Volume 9, Southwest: Washington, D.C., Smithsonian Institution, 701 p.

Palmer, Tim, 1991, The Snake River: Washington, D.C., Island Press, 321 p.

Partridge, J., and Baker, V.R., 1987, Paleoflood hydrology of the Salt River, Arizona: Earth Surface Processes and Landforms, v. 12, p. 109-125.

Petts, G.E., 1979, Complex response of river channel morphology subsequent to reservoir construction: Progress in Physical Geography, v. 3, no. 3, p. 329-362.

Petts, G.E., 1984, Impounded Rivers, Perspectives for Ecological Management: New York, John Wiley and Sons, 326 p.

Pickup, Geoff, 1976, Adjustment of stream-channel shape to hydrologic regime: Journal of Hydrology, v. 30, p. 365-373.

Powell, J.W., 1879, Report on the Lands of the Arid Region of the United States, with a More Detailed Account of the Lands of Utah (2nd edition): Washington, D.C., U.S. Government Printing Office, 195 p.

Pucherelli, M.J., Clark, R.C., and Williams, R.D., 1990, Mapping backwater habitat on the Green River as related to the operation of Flaming Gorge Dam using remote sensing and GIS: Bureau of Reclamation Report R-90-18, 11 p.

Randle, T.J., Strand, R.I., and Streifel, A., 1993, Engineering and environmental considerations of Grand Canyon sediment management, *in* Engineering Solutions to Environmental Challenges: Chattanooga, Tennessee, U.S. Committee on Large Dams, Thirteenth Annual USCOLD Lecture.

Reinecke, K.J., and Krapu, G.L., 1986, Feeding ecology of sandhill cranes during spring migration in Nebraska: Journal of Wildlife Management, v. 50, no. 1, p. 71-79.

Reinhardt, G.F., 1937, Rectification of the Rio Grande in the El Paso/Juarez Valley: American Journal of International Law, v. 31, p. 44-54.

Robinson, T.W., 1965, Introduction, spread, and areal extent of saltcedar *(Tamarix)* in the western states: U.S. Geological Survey Professional Paper 491-A, 12 p.

Schefter, J.E., and Hirsch, R.M., 1980, An economic analysis of selected strategies for dissolved-oxygen management— Chattahoochee River, Georgia: U.S. Geological Survey Professional Paper 1140, 29 p.

Schmidt, J.C., 1990, Recirculating flow and sedimentation in the Colorado River in Grand Canyon, Arizona: Journal of Geology, v. 98, p. 709-724.

Schmidt, J.C., 1992, Temporal and spatial changes in sediment storage in Grand Canyon: Flagstaff, Arizona, U.S. Department of the Interior, National Park Service Cooperative Agreement #CA 8006-8-0002, 21 p.

Schmidt, J.C., and Graf, J.B., 1990, Aggradation and degradation of alluvial sand deposits, 1965 to 1986, Colorado River, Grand Canyon National Park, Arizona: U.S. Geological Survey Professional Paper 1493, 74 p.

Schmidt, J.C., Webb, R.H., and Grams, P.E., 1995, Comparison of the magnitude of erosion of alluvial bars along two large regulated rivers: Water Resources Bulletin, v. 31, p. 617-631.

Smillie, G.M., Jackson, W.L., and Tucker, D., 1992, Colorado River sand budget—Lees Ferry to Little Colorado River including Marble Canyon: Flagstaff, Arizona, U.S. Department of the Interior, National Park Service Cooperative Agreement #CA 8006-8-0002, 8 p.

Smith, N., 1971, A History of Dams: Secaucus, New Jersey, Citadel Press, 279 p.

Stevens, L.E., and Ayers, T.J., 1993, Impacts of Glen Canyon Dam on riparian vegetation and soil stability in the Colorado River corridor, Grand Canyon, Arizona: Flagstaff, Arizona, U.S. Department of the Interior, National Park Service Cooperative Agreement #CA 8000-8-0002, p. 2/1 to 2/5.

Stokes, W.R., McFarlane, R.D., and Buell, G.R., 1991, Water resources data, Georgia, Water year 1991: U.S. Geological Survey Water-Data Report GA-91-1, 540 p.

Stuebner, Stephen, 1993, Salmon advocates say—the quiet slaughter continues: High Country News, v. 25, n. 10, p. 1.

Thomas, H.H., 1976, The Engineering of Large Dams: London, John Wiley and Sons, 777 p.

Tyus, H.M., 1990, Potamodromy and reproduction of Colorado squawfish in the Green River Basin, Utah, 1979-1986: Transactions of the American Fisheries Society, v. 119, p. 1035-1047.

Tyus, H.M., and Karp, C.A., 1991, Habitat use and streamflow seeds of rare and endangered fishes in Green River: Vernal, Utah, Flaming Gorge Studies Program, U.S. Fish and Wildlife Service, July 31, 1991, variable page numbering.

U.S. Department of the Interior, 1981, Water and Power Resources Service, Project Data: Washington, D.C., Water and Power Resources Service, 1463 p.

U.S. Department of the Interior, 1989, Glen Canyon Environmental Studies, Final report: Salt Lake City, Utah, Bureau of Reclamation, Upper Colorado Region, NTIS Report No. PB-88-183348/AS, 84 p.

U.S. Department of the Interior, 1995, Operation of Glen Canyon Dam, Final environmental impact statement: Salt Lake City, Utah, Bureau of Reclamation, Upper Colorado Region, 337 p.

Valdez, R.A., 1989, Winter habitat study of endangered fish—Green River: Logan, Utah, Bio/West Report Number 136-2, Bureau of Reclamation Contract No 6-CS-40-04490, 178 p.

Valdez, R.A., 1990, The endangered fish of Cataract Canyon: Logan, Utah, Bio/West Report No 134-3, Bureau of Reclamation Contract Number 6-CS-40-03980, 94 p.

Valdez, R.A., 1993, A fish with finesse: Grand Canyon River Guides News, v. 6, no. 1, p. 12-14.

Vanicek, C.D., and Kramer, R.H., 1969, Life history of the Colorado Squawfish and Colorado Chub in the Green River in Dinosaur National Monument, 1964-1966: Transactions of the American Fisheries Society, v. 98, no. 2, p. 193-208.

Vanicek, C.D., Kramer, R.H., and Franklin, D.R., 1970, Distribution of Green River fishes in Utah and Colorado following closure of Flaming Gorge Dam: Southwestern Naturalist, v. 14, no. 3, p. 297-315.

Webb, R.H., 1996, Grand Canyon, A Century of Change: Tucson, University of Arizona Press, 320 p.

Webb, R.H., Pringle, P.T., and Rink, G.R., 1989, Debris flows from tributaries of the Colorado River, Grand Canyon National Park, Arizona: U.S. Geological Survey Professional Paper 1492, 39 p.

Wegner, D.L., 1991, A brief history of the Glen Canyon Environmental Studies, *in* Colorado River Ecology and Dam Management: Washington, D.C., National Academy Press, p. 226-238.

Williams, G.P., 1978, The case of the shrinking channels—the North Platte and Platte Rivers in Nebraska: U.S. Geological Survey Circular 781, 48 p.

Williams, G.P., and Wolman, M.G., 1984, Downstream effects of dams on alluvial rivers: U.S. Geological Survey Professional Paper 1286, 83 p.

Wilson, R.P., 1986, Sonar patterns of Colorado riverbed, *in* Proceedings of the Fourth Federal Interagency Sedimentation Conference, Volume 2: Las Vegas, p. 5-133 to 5-142.